**Food Safety in the Seafood I**

*Para as Filipas e Pais*

# Food Safety in the Seafood Industry

A practical guide for
ISO 22000 and FSSC 22000
implementation

**Nuno F. Soares**

Vanibru, Lda.
Braga, Portugal

**Cristina M. A. Martins**

SGS Portugal Group
Maia, Portugal

**António A. Vicente**

Associate Professor with Habilitation
Department of Biological Engineering
University of Minho
Braga, Portugal

WILEY Blackwell

*Library of Congress Cataloging-in-Publication Data*

Names: Soares, Nuno F., 1975– author. | Martins, Cristina M. A., 1990– author. |
    Vicente, António A., author.
Title: Food safety in the seafood industry : a practical guide for ISO 22000 and FSSC 22000
    implementation / Nuno F. Soares, Cristina M. A. Martins, António A. Vicente.
Description: Chichester, UK ; Hoboken, NJ : John Wiley & Sons, 2016. | Includes index.
Identifiers: LCCN 2015040940 (print) | LCCN 2015044582 (ebook) | ISBN 9781118965078 (pbk.) |
    ISBN 9781118965085 (ePub) | ISBN 9781118965092 (Adobe PDF)
Subjects: LCSH: Seafood industry–Quality control. | Seafood industry–Safety measures. |
    Seafood–Quality control. | Seafood–Safety measures.
Classification: LCC HD9450.5 .S63 2016 (print) | LCC HD9450.5 (ebook) | DDC 363.19/29–dc23
LC record available at http://lccn.loc.gov/2015040940

A catalogue record for this book is available from the British Library.

# Contents

# Foreword

In May 2014 I was contacted by one of the authors of this book, Nuno Soares (unknown colleague until that day), who asked to meet with me to present the idea of writing a new book. In our first meeting I heard his enthusiastic words and, as I do with almost all the people who contact me, I shared with Nuno some literature sources and several personal contacts. It seemed to me excellent that the authors of a book had, as a priority, the personal contact with experts in the topics covered and intended to collect, by live interviews, their testimonies.

About a year later Nuno contacted me again with two surprises: to inform me that the book was already in the phase of last corrections and to invite me to write the foreword. I felt honored by the invitation and accepted this challenge with pleasure, but also some apprehension. Pleasure because I always liked books, writing, and communication of ideas; apprehension because it does not seem easy to write an interesting foreword which contributes positively to the value of an international book, especially such a technical and practical book as that which the reader is holding.

This book covers recent management aspects of safety of seafood. Seafood is, without any doubt, one of the most important food groups that have a positive influence on human health (consequently, seafood is one of the food items of the future) and, simultaneously, one that includes the most diverse and most perishable foods. As a result, safety issues that seafood poses during its vast and complex production, processing, and distribution chains are very specific, very delicate, and require a multifaceted and complex approach and management.

In the seafood sector, as much or even more than in other food groups, the development, dissemination, and compliance with international standards has been undoubtedly one of the most effective ways to improve safety, in order to move it towards the theoretical maximum levels that are considered unattainable. At a very fast speed (I would say frightening), new international rules have been published. Despite the fact that many of these rules are mandatory, it is very difficult for industry professionals to be informed about all of them and to adapt themselves and their industry rapidly and adequately in order to work strictly within legal limits.

It is precisely for this particular reason that I believe this book is unique and very useful. Using simple and accessible language, supported by tables and images showing clear facts and figures, the reader is introduced to the current context of the seafood characteristics and its specific safety issues in the first two chapters. In the following chapters, the main current rules applicable to the fish sector are examined in a concise and practical way; the objective of the authors is always to make it easy for readers to understand and implement these rules. The authors also present insights about costs and benefits associated with the implementation of the covered standards, which greatly help managers. Without a clear understanding of the benefits to be gained, it is not easy to approve costs; benefits can be high, especially over the long term.

This book also contains testimonials from world-class experts in the topics discussed, interviewed expressly for this purpose, exploring the historical perspective and other details of the standards covered. For me, this is what makes this book significantly different from many other publications. The direct contact with experts from around the world (taking advantage of new forms of electronic communication) is irreplaceable and brings unique insights; the sharing of ideas and experiences and the discussion of problems and solutions, which would be more difficult by simple analysis of written elements, is invaluable. This book offers readers both possibilities: the reading of a clear text, and the contact with world-renowned experts on the covered themes.

Another interesting characteristic of this book is the appendix with a practical analysis of data collected directly from the Rapid Alert System of Food and Feed (RASFF), followed by identification of the real dangers and risks posed by seafood. This real-data-approach appendix is a very useful complement to the theoretical information presented in the main part of the book.

Readers therefore have in their hands a valuable aid that they can use in two ways: (1) as a reading book for general information and knowledge about the rules and their practical application, particularly suitable at an early stage of study and for a more theoretical approach; and also (2) as a rapid reference book, to answer questions posed during practical implementation and subsequent phases.

Since the definition of 'foreword' includes the word 'brief', I would like to end by congratulating the authors for the excellent idea, for the courage of the choice of a paper edition, and for the production of a useful and valuable book. I hope that our colleagues in the seafood area, all over the world, buy and use this working element, and thereby contribute to its commercial success. I am sure that the purchase of the book is a very small investment compared to the great benefit it certainly represents to its readers.

Happy reading and success to all.

Paulo Vaz-Pires
Marine Biologist, Professor and researcher on Seafood Quality and Safety at ICBAS, University of Porto, Portugal.
E-mail: vazpires@icbas.up.pt

# Acknowledgements

The authors are grateful for the collaboration of Paulo Vaz-Pires (Associate Professor at Instituto Ciências Biomédicas Abel Salazar, Univ. Porto), Elisabete Nunes (Propeixe), Tânia Mendes, Pedro Silva, Zita Avelar, and Juliana Pereira. Special thanks are due to Lone Skjernin (Secretariat at ISO/TC 34/SC 17), Cornelie Glerum (Secretary general at Food Safety System Certification 22000), Lisa Prévert (Communications Officer at Global Food Safety Initiative & Global Social Compliance Programme), and Ana Silva for support, collaboration, and information. We are indebted to Tom Ross (Associated Professor at the University of Tasmania), Bill Marler (Food safety advocate at Marler Clark), and William Sperber (Cargill's Global Ambassador for Food Protection) for their personal insights, which undoubtedly enriched this book beyond what we could have anticipated.

## Disclaimer

The content of this book integrates the research and findings of the authors during the period 2014–2015 and their personal experience, and is provided without any liability whatsoever in its application and use. It is not intended to represent the view of Vanibru – Comércio de Produtos Alimentares Lda, SGS Portugal, or Universidade do Minho.

# Introduction

Most professionals in the seafood industry, even those with experience in quality management or education in the field, find it difficult to implement Food Safety Systems. Most companies need to recruit consultants to find this type of expertise, and it is often difficult to find someone with experience in this particular field. This book will provide a hands-on approach to the understanding and implementation of the standard ISO 22000 and the additional requirements needed to comply with the Food Safety System Certification 22000 (FSSC 22000) approach for ISO-based certification schemes. It will also briefly characterize the seafood industry and how ISO 22000 and FSSC 22000 can be important and valuable tools in its future. The objective was to provide, in a single document, fundamental information key to the understanding of ISO 22000 and FSSC 22000 and assistance to anyone who is implementing them, using non-technical language which is easily understandable by non-specialists while also being useful for food safety technicians. The book aims to be geographically global but industry specific.

The book is structured to be clear and assertive in key points, a powerful tool and a useful reference for professionals in the field, company managers, consultants, auditors, teachers, and students.

The book deals with three main subjects – seafood, ISO 22000, and FSSC 22000 – in five separate chapters. Chapter 1 presents a brief characterization of seafood and the seafood industry, followed by the identification and description of the most relevant hazards for seafood products. Food safety is the focus of Chapter 2 where, as well as an introduction to the *Codex Alimentarius* and hazard analysis and critical control points (HACCP; including the most fascinating contribution of Dr Sperber with some insights into the early days of HACCP), there is also a brief introduction to three of the most-used Food Safety Systems (BRC Global Standard for Food Safety, SQF Code, and IFS Food Standard) that, together with FSSC 22000, are Global Food Safety Initiative (GFSI) benchmarked.

In Chapter 3 ISO 22000 is introduced together with the challenges and drivers of its implementation. Chapter 3 also includes an interesting interview with William Marler, the most prominent foodborne illness lawyer in America, who describes the importance of food safety systems and third-party audits not only in terms of food safety but also as a valuable management defense if food safety issues arise.

All the clauses of ISO 22000:2005 are scrutinized and explained in Chapter 4 with the focus on how organizations can address them, considering both the approach of *ISO 22004: Guidance on the application of ISO 22000* (2014) and the personal experience of the authors.

Chapter 5 introduces FSSC 22000, its history and the additional requirements to ISO 22000:2005 that must be fulfilled to obtain certification.

The book has two appendices. The first is a comprehensive list of documents that need to be implemented and maintained according to ISO 22004:2014. The second is a compilation of the hazards notified to Food and Feed Safety Alerts (RASFF) related to seafood products, since this database was created by the European Commission.

This book is not an end in itself. It is a tool and a vehicle for further cooperation and information interchange related to seafood safety and food safety systems. QR codes can be found throughout the book; when scanned they will allow the reader to contact the authors directly, know their personal views on each chapter and even access or request more details on the book content. We strongly encourage the readers to use the QR codes or contact us through foodsafetybooks@gmail. com with comments, suggestions, or questions.

# CHAPTER 1

# Fishery sector

## 1.1 Characterization of seafood

### 1.1.1 Classification

The term 'seafood' used throughout this book represents three categories of organisms – fish, crustaceans, and mollusks – each of them belonging to a different phylum within the kingdom Animalia.

Identification of fish from different species by nonproperly trained people can be very challenging and even impossible most of the times. The use of local or common names can also originate misunderstandings; the same species may have distinct names in different regions or the same name may be attributed to different species. The best way to avoid such mistakes is the use of the scientific name (in Latin) to clearly identify seafood species all over the world. This clarification is also of great importance since the economic value of seafood can be dependent on the species.

In taxonomic terms, the majority of commercially relevant fish species category belong to the phylum Chordata (subphylum Vertebrata), which is divided into different classes among which stands out the class of ray-finned fish Actinopterygii (superclass Osteichthyes, also called bony fish) (Nelson 2006; Auerbach 2011). By the fact that their skeleton is entirely composed of cartilage, sharks, rays, and skates belong to the class of cartilaginous fish Chondrichtyes (Huss 1988; Auerbach 2011).

Crustaceans belong to the phylum Arthropoda and to the subphylum Crustacea. Within this subphylum, the class Malacostraca stands out for being the class that has the largest number of known species by far (Saxena 2005; Auerbach 2011). This class includes shrimps, prawns, crabs, and lobsters which, in turn, constitute the order Decapoda (Saxena 2005).

*Food Safety in the Seafood Industry: A Practical Guide for ISO 22000 and FSSC 22000 Implementation,*
First Edition. Nuno F. Soares, Cristina M. A. Martins and António A. Vicente.
© 2016 John Wiley & Sons, Ltd. Published 2016 by John Wiley & Sons, Ltd.

Finally, mollusks belong to the phylum Mollusca, which is divided into several classes. Bivalve mollusks, such as mussels, oysters, scallops, and clams, belong to the class Bivalvia (also known as Lamellibranchia or Pelecypoda), and cephalopod mollusks (e.g., squids, octopuses, and cuttlefishes) belong to the class Cephalopoda or Siphonopoda (Haszprunar 2001; Helm *et al.* 2004; Auerbach 2011).

## 1.1.2 Anatomy
### Bony fish
The skeleton of bony fish, as the name suggests, is totally made of bones. Wheeler & Jones (1989) suggested that the skeletal structure of bony fish could be divided into two parts: head skeleton and axial skeleton. The head skeleton is composed of three systems: (1) neurocranium, which surrounds and protects the brain and the sense organs; (2) bones system, that is related to feeding; and (3) combined hyal and branchial systems, which form gill arches and gill covers. The axial skeleton is formed of a set of articulated vertebrae that range from head to tail forming the vertebral column or backbone (Huss 1988; Wheeler & Jones 1989).

According to Schultz (2004), the body of bony fish has three types of muscles: smooth, cardiac, and striated (edible part). Although most fish muscle tissue is white, certain species (e.g., pelagic fish, such as herring and mackerel) have a portion of reddish- or brown-colored tissue. The so-called dark muscle is located under the skin or near the spine (Huss 1988, 1995). According to Love (1970), fish activity causes variations on the proportion of dark to white muscle. For instance, the dark muscle of pelagic fish (i.e., species which swim more or less continuously) could represent up to 48% of the body weight. The chemical composition of dark muscle differs from that of white muscle since it contains higher amounts of lipids, myoglobin, alkali soluble proteins, stroma, and glucogen (Chaijan *et al.* 2004; Bae *et al.* 2011). These differences, especially the high lipid content found in the dark muscle, are directly responsible for problems related to rancidity (Huss 1988). Moreover, muscle composition is relevant in terms of ability to cause an allergic reaction. A glycoprotein named parvalbumin, which is responsible for triggering the immune response leading to allergy symptoms, has been demonstrated to be 4–8 times higher in white muscle compared to dark muscle (Kobayashi *et al.* 2006).

Bony fish have a skin which is commonly covered by scales and they use the gills for breathing underwater, as seen in Figure 1.1. There are different organs within the fish body which form part of the digestive system including stomach, intestine, and liver, which are commonly known as guts (Johnston *et al.* 1994). Because many pathogenic bacteria are commonly present in the normal gut microflora, evisceration is the first critical step to control contamination of fish flesh after handling and before freezing.

### Crustaceans
Crustaceans are classified as arthropods and are characterized by the presence of a hard exoskeleton made of chitin and a segmented body with appendages on each segment (Adachi & Hirata 2011).

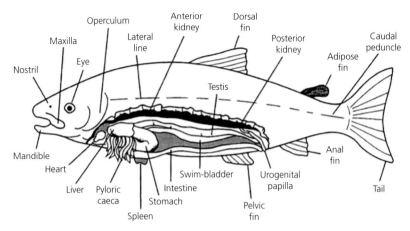

**Figure 1.1** Diagram of the basic anatomy of a salmonid fish. Source: Roberts (2012). Reproduced with permission from John Wiley & Sons.

According to Raven & Johnson (2002), most species belonging to the Sub-phylum *Crustaceae* have two pairs of antennae, three types of chewing appendages, and a different number of legs, as presented in Figure 1.2. Shrimps, prawns, crabs, and lobsters, which are a very important fishery resource, have ten legs in the form of thoracic appendages. This characteristic reflects the origin of the name *Decapoda*, a word that derives from the Greek words for ten (*deka*) and feet (*pous*) (Ng 1998). The carapace of decapod crustaceans is reinforced with calcium carbonate and their head and thoracic segments are fused, forming a structure called cephalothorax. These animals can have a telson (or tail spin) in the terminal region of the body (Raven & Johnson 2002).

Allergies to crustaceans are common and usually more publicized than allergies to other seafood products. Tropomyosin, a water soluble and heat-stable muscle protein, has been identified as the major allergen of shrimp (Shanti *et al.* 1993; Daul *et al.* 1994). Tropomyosin can also be responsible for allergic reactions in other products such as mollusks, but it has not been demonstrated that this allergen cross-reacts with fish allergens (Lopata *et al.* 2010).

## Bivalve mollusks

Bivalve mollusks such as mussels, oysters, scallops, and clams are invertebrates characterized by the presence of a shell. According to Gosling (2003), the shells of bivalve mollusks are mainly formed of calcium carbonate in three different layers: first, an inner calcareous (nacreous) layer; second, an intermediate layer of aragonite or calcite; and finally a thin outer periostracum of horny conchiolin. Depending on the species, shells can have a variety of shapes, colors and markings. For that reason, the characteristics of shells are commonly used in the identification of diverse species of bivalves (Poutiers 1998; Gosling 2003).

The shells of mollusks shells are formed of two valves, which laterally compress their soft body, and are dorsally hinged by an elastic and poorly calcified structure,

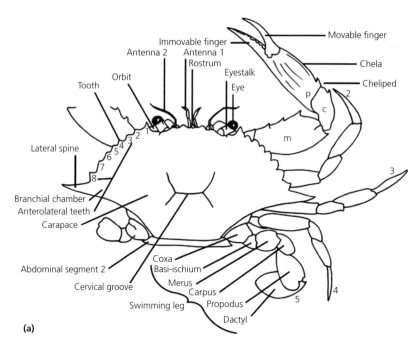

**(a)**

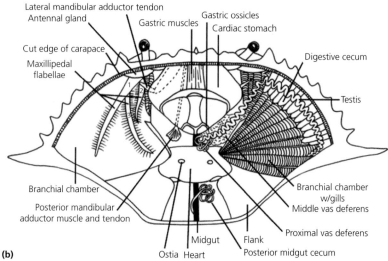

**(b)**

**Figure 1.2** Schematic drawing of a male blue crab. (a) Dorsal view of external anatomical features. (b) Dorsal view of internal anatomy. Source: Lewbart (2011). Reproduced with permission from John Wiley & Sons.

the ligament (Poutiers 1998; Helm *et al.* 2004). This ligament is also involved in the system that controls the opening and closing of both valves. Shells are closed due to the contraction of the adductor muscle(s), which causes a reaction of stretching within the ligament. When the muscle relaxes, the ligament tends to contract and releases the created tension. As the ligament returns to its initial position the valves depart from one another, hence opening the shell (Ray 2008).

According to Helm *et al.* (2004), when one of the valves is removed it is possible to see the internal organs of the mollusks. All the organs are covered by a mantle, shown in Figure 1.3, among which the gills or ctenidia stand out. This organ is

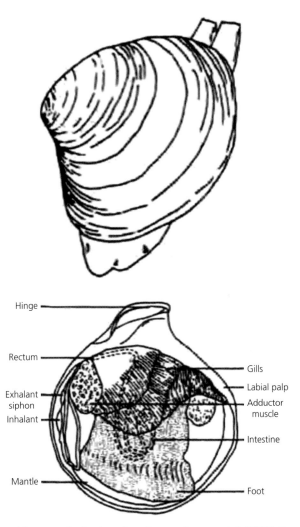

**Figure 1.3** Internal features of a Quahog clam. Source: Granata *et al.* (2012). Reproduced with permission from John Wiley & Sons.

used to filter food from water and to breathe (Coan & Valentich-Scott 2006). However, some contaminants present in the environment, such as pathogenic bacteria, viruses, and chemicals, are commonly retained inside.

## Cephalopod mollusks

As mentioned above, cephalopod mollusks, which include squid, octopus, cuttle-fish, and nautilus, belong to the class Cephalopoda, one of the major and most complex classes of the phylum Mollusca (Jereb & Roper 2005; Ray 2008).

The name *Cephalopoda* derives from the combination of two Greek works: *kefale* and *pous* which mean head and feet, respectively. This is related to the fact that the members of this class have a head that supports a set of arms or both arms and tentacles, placed in a circle around its mouth. These appendages (arms or ten-tacles) are provided with many suckers or hooks, helping them to capture and hold prey (Jereb & Roper 2005). Like other mollusks (e.g., bivalves and gastro-pods), cephalopods have an external shell. However, according to several authors (Boyle & Rodhouse 2005; Jereb & Roper 2005), the greater part of the living forms of these animals lost their shell or it was reduced. For instance, in squids and cuttlefish the external exoskeleton was reduced; they presently possess an internal shell called gladius, pen, or cuttlebone. An outer shell for protection is only present in the living cephalopods from the Family *Nautilidae* (Dunning & Wadley 1998). Jereb & Roper (2005) concluded that the loss of the external shell allowed the development of a muscular mantle which covers the internal organs (Jereb & Roper 2005). Figure 1.4 depicts the basic features of a squid.

### 1.1.3 Chemical composition

When the subject is food safety, it is fundamental to know the chemical composi-tion of seafood. With this knowledge, professionals are able to foresee what kind of microorganisms can develop in seafood and which changes may occur in the product after harvesting and during shelf life.

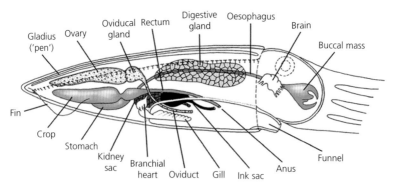

**Figure 1.4** Generalized anatomy of a loliginid squid. Source: Boyle & Rodhouse (2007). Reproduced with permission from John Wiley & Sons.

According to Murray & Burt (2001), the chemical constituents of fish flesh can be divided into two groups: the major and minor components. The former comprises water, protein, and fat, whereas the latter includes carbohydrates, minerals, and vitamins. The amount of each constituent can be influenced by extrinsic factors (e.g., the environment/season) or by intrinsic factors of the fish (e.g., species, age, sex, or spawning/migration period) (Huss 1988, 1995).

## Water

Water is well known as a fundamental substance to maintain life on Earth and is the main constituent of all living organisms. According to Murray & Burt (2001), water typically comprises up to 80% of a lean fish fillet weight and about 70% of a fatty fish flesh weight. However, these values may vary between 30% and 90% in certain species. Several authors (Feeley *et al.* 1972; Love 1980, 1988; Huss 1988; Osman *et al.* 2001; EFSA 2005; Pirestani *et al.* 2009) reported that water content in fish varies inversely to the fat percentage, that is, water content is higher in low-fat species than in fatty fish. Regarding the three categories described in Section 1.1.1, mollusks have more water than fish and crustaceans.

The absolute content of water present in food usually takes two forms: (1) free or available water; and (2) bound or unavailable water. Water that is not linked to any component, such as proteins or carbohydrates, is available for growth of microorganisms including pathogenic bacteria (Dauthy 1995). According to Jay *et al.* (2005), the amount of water available for microbial growth is described in terms of water activity ($a_w$) and can vary between 0 and 1 (Neumeyer *et al.* 1997; Aberoumand 2010). Jay *et al.* (2005) reported that the value of this parameter exceeds 0.99 in the majority of fresh foods and, according to Martin *et al.* (2000), fresh fish has a water activity close to 1, making it vulnerable to contamination.

Each microorganism has a different water activity range: pathogenic and spoilage bacteria require a high amount of water and do not grow in foods with a water activity of less than 0.85, whereas many yeasts and moulds can grow at water activity values as low as 0.60 as shown in Table 1.1 (Jay *et al.* 2005; FDA 2011). In order to prevent microbial growth, there are a number of strategies that can be applied, namely freezing, drying, and addition of solutes or ions.

## Proteins

Proteins are chains of small units called amino acids linked to one another to make a long molecule. There are 20 different naturally occurring amino acids, most of them essential for the maintenance of good health, making their presence in the human diet very important. A healthy human diet should include the ingestion of amino acids in balanced proportions. A proper combination of amino acids to meet the nutritional needs of man can be provided by fish protein equally as supplied by meat, milk, and eggs (Murray & Burt 2001).

Fish protein is easily digested and has a high biological value (Bohl *et al.* 1999; EFSA 2005). Among species, the amino acid content of fish meat is similar. The protein content of their edible parts is similar to the muscle meat

**Table 1.1** Approximate minimum $a_w$ values for growth of microorganisms important in foods

| Microorganism | Minimum $a_w$ (using salt) (FDA 2011) | Minimum $a_w$ (using salt) (Jay et al. 2005) |
|---|---|---|
| Bacillus cereus | 0.92 | – |
| Campylobacter jejuni | 0.987 | – |
| Clostridium botulinum (type A) | 0.935 | 0.94 |
| Clostridium botulinum (type E) | 0.97 | 0.97 |
| Clostridium perfringens | 0.93 | – |
| Escherichia coli | 0.95 | 0.96 |
| Listeria monocytogenes | 0.92 | – |
| Salmonella spp. | 0.94 | – |
| Shigella spp. | 0.96 | – |
| Staphylococcus aureus (toxin production) | 0.83 | 0.86 |
| Vibrio cholerae | 0.97 | – |
| V. parahaemolyticus | 0.94 | 0.94 |
| V. vulnificus | 0.96 | – |
| Yersinia enterocolitica | 0.945 | – |

of animals but, in contrast to cuts from many animals, the uniformity/homogeneity of fish is higher (EFSA 2005).

The amount of protein present in fish muscle is around 15–20%; however, values as high as 28% can be found in some species of fish (Murray & Burt 2001). According to EFSA (2005), the protein content decreases somewhat with age-related increases in the lipid content, despite the fact of being similar in fish on a weight basis (15–20 g/100 g).

## Lipids

Often referred to as fats, lipids include fats, oils, waxes, and other compounds of fatty acids (Murray & Burt 2001). Commonly, fish is divided according to the fat percentage of body weight. Fatty fish (5–20%) accumulates fat in muscle tissue and lean fish (1–2%) accumulates fat predominantly in the liver (EFSA 2005). The lipid content of fish varies not only between different species but also within the same species according to season and feeding. Lipids are unevenly distributed even within a particular individual; in salmon for example, near the head the lipid content is double that in tail muscle (Murray & Burt 2001).

Similarly to most vertebrates, in most fish species fat depots are composed of triglycerides. However, fish lipids are different from mammalian lipids since they are composed of long-chain polyunsaturated fatty acids (LC-PUFAs) containing many fatty acids with five or six double bonds (Stansby & Hall 1967; Huss 1988). These LC-PUFAs, especially eicosapentaenoic and docosahexaenoic acids, are very important in human nutrition since they cannot be synthesized (due to the absence of the enzyme that synthesizes alpha-linoleic acid). They are conventionally known as ω-3, indicating that the first double bound is located after the third carbon atom from the methyl end of the chain. LC-PUFAs are associated with important functions such as brain development in children, in the last trimester of pregnancy, and disease prevention such as sudden cardiac death, coronary heart disease, and atherosclerosis. Another benefit of seafood is that its consumption can produce effects very rapidly, within weeks or months, as in the case of lowering blood pressure or anti-thrombotic actions (FAO 2013). Most of these benefits have been known of since the middle 1970s and early 1980s when Danish scientists acknowledged that Greenland Eskimos, despite the large presence of fat and cholesterol from marine foods in their diet, rarely suffer from ischemic heart disease and have lower low-density lipoprotein (LPL), cholesterol, and triglyceride levels than Denmark Eskimos (typically on a western European diet). Although there is convincing evidence of fish LC-PUFAs benefits, it is a great challenge to define a dose–response relationship; not only does the concentration of these constituents differ among distinct products, but the benefits obtained may not be linearly dependent on the amount consumed (Mozaffarian & Rimm 2006).

## Carbohydrates

Carbohydrates are substances containing only elements of carbon, hydrogen, and oxygen that, when combusted, produce carbon dioxide and water. This group of substances includes sugars, starches, gums, and celluloses. Fish is a poor source of carbohydrates. Generally, the amount of carbohydrates in fish muscle is less than 1%, although other tissues such as liver can provide higher values. In the dark muscle of some fatty species and in some mollusks carbohydrate content can increase up to 2% or 5%, respectively (Aitken et al. 2001; Murray & Burt 2001).

## Minerals and vitamins

Seafood is a good source of minerals and vitamins for a healthy diet. Usually reported in compositional tables as ashes, because of their quantification method, minerals are inorganic compounds that typically represent less than 2% of the edible portion (Aitken et al. 2001) of fish. For example, iodine and selenium are almost exclusively found in foods from the aquatic environment and are important for the development of brain and neural systems of children (Toppe 2012). Iodine deficiency is estimated to affect about one-third of the global human population and can lead to brain damage and mental retardation (de Benoist et al. 2008). According to EFSA (2005) the amount of iodine present in fish has been reported to range from 5 μg to 1210 μg iodine per 100 g fresh weight of edible parts. According to the same report, it is also possible to observe that all species of fish contain considerable amounts of selenium. This element presents a high binding affinity for mercury and can therefore reduce mercury toxicity (Ralston et al. 2008).

Vitamins can be divided into two groups, namely: those that are soluble in fat, such as vitamins A, D, E, and K; and those that are soluble in water, such as vitamins B and C. One of the main characteristics of the vitamin content in fish muscle is that it varies significantly between different fish species and even between some parts of the same fish. The vitamin content of fish muscle is comparable to that of mammals and in general is a great source of fat-soluble vitamins A, D, and E (Huss 1988). Vitamin D is particularly important since it is not found in many foods and is important to bone health and type-2 diabetes reduction, among other benefits. Despite the fact that vitamin D can be produced from skin exposure to sunlight, some populations can suffer from its deficiency due to long winters or the use of clothes that cover the entire body. Water-soluble vitamins, such as B and C, tend to be more uniformly distributed through the flesh (Murray & Burt 2001). Vitamin B has an important role in energy production in the cells. Since vitamins can be sensitive to factors such as light, heat, temperature, and storage time, processing and storage conditions are important to preserve the natural vitamin content of fish.

### 1.1.4 Marine ecosystem

The management of fisheries and marine ecosystems is critical to reduce the increasing degradation and loss of marine habitats, to improve research and define policies for marine science, and to manage conflicts in the use of marine resources. The need to exercise some control over the various uses of a maritime area became evident in the late twentieth century as a result of concern for the health of oceans and the need to regulate human activities, which were generating an increasing occurrence of conflicts.

The proper functioning of an ecosystem results from the capacity of a community to adapt to the physical environment and its relationships with other communities, consisting of populations of species that have their own dynamics in terms of abundance, survival, growth, production, reproduction, and other strategies.

The geographic boundaries of ecosystems are often difficult to define, since the extent, location, and structure of an ecosystem, as well as species composition and functioning, may vary seasonally or change every year under certain climatic conditions.

The ocean as an ecosystem is not a uniform and constant environment. Weather phenomena such as *El Niño* alter the oceans, both physically and biologically, and also affect the distribution of fish. In addition to these changes in oceanographic features, other changes occur as a result of events such as *El Niño* such as in sea surface temperature, vertical thermal structure of the ocean (particularly in coastal regions), coastal and upwelling currents, and patterns of migration. All these changes directly affect species composition and abundance of fish, while fish that remain in the affected regions experience a reduction in growth, reproduction, and survival. Fluctuations in fish stocks are therefore not exclusively caused by fishing, but also by environmental conditions. *El Niño*-induced weather changes are felt worldwide, with important implications in the dynamics of aquatic ecosystems. According to Tudhope *et al.* (2001) and Garcia *et al.* (2004), *El Niño*-like conditions are becoming more frequent and more intense, making a deeper understanding of the impact of those events in the ecosystems extremely important (Garcia *et al.* 2004).

An extensive knowledge of the characteristics of ecosystems, namely complexity, structure, functioning, natural variability, and boundaries, as well as the implementation of better management techniques are essential for maintaining the productivity of ecosystems for present and future generations. To achieve that goal, it is imperative to maintain habitats, reduce pollution and degradation, minimize waste, and protect threatened and endangered species.

The closer ecosystems are to human intervention, the greater the impact of pollution on them. In inland or coastal waters, which are more susceptible, direct intervention with the habitat can minimize environmental problems, while in marine ecosystems, due to the fact that direct intervention is limited, the focus is mainly on the control of certain human activities (namely fishing).

The complexity of the interactions between different marine species, either in the predator/prey relationship or between species of the same trophic level, makes the prediction of the consequences of perturbations introduced in the ecosystem very difficult. For example, diseases or contaminants that are introduced in a trophic level will be transmitted to their predators by feeding. The fishing of certain species and consequent reduction of its abundance in the ecosystem also has consequences not only on their predators but also on their prey and other species that compete for the same food sources.

Because we cannot fully understand the ecosystem structure and functioning and as it is extremely difficult to distinguish between natural and human-induced changes, most of the consequences of human interventions are not always perfectly predictable and/or reversible. However, it is well known that fishing affects the ecological processes on a large scale and that overfishing can transform an originally stable, mature, and efficient ecosystem into an immature and stressed ecosystem (Garcia *et al.* 2003).

## 1.2  Characterization of the seafood industry

### 1.2.1  Development of the fish industry

The fish industry has evolved significantly over the last few decades. This development was driven by several factors such as population growth, expansion of production and processing of fish, greater concern of consumers with a healthy diet, and increasing efficiency of distribution channels. The combination of all these factors with a growing interest in food safety and quality has forced the adoption of increasingly stringent hygiene actions to protect the health of consumers. Given the high perishability of fish, all steps of handling, processing, preservation, packaging, and storage become essential to improve its shelf life, ensure safety, maintain quality and nutritional attributes, and avoid waste and losses.

Currently, the fish industry processes seafood in a large variety of products and uses different packaging materials and conditions. This transformation was largely due to technological developments in recent decades related to food processing and packaging and to an increasing importance given to product differentiation. For instance, innovations in refrigeration, ice making, packing, and shipping allowed product integrity to be ensured and fish distribution to be expanded in different forms such as fresh, chilled, and frozen.

In developing countries, such advances in processing have not yet occurred and sophisticated methods of processing, such as filleting, salting, canning, drying, and fermentation, are still not used or are used to a lesser extent due to a lack of infrastructure and services (including hygienic landing centers, electricity, drinking water, roads, ice, cold rooms, and appropriate refrigerated transport). However, a few signs of evolution have been observed. For instance, frozen fish production for human consumption has increased from 13% to 24% between 1992 and 2012 (FAO 2014).

## 1.2.2  Fish consumption and international trade

Fish utilization can be categorized as 'nonfood' or 'food' use. Since the beginning of 1990, the proportion of fish production used directly for human consumption has increased. In fact, in the 1980s about 71% of fish produced was intended for human consumption, increasing up to 73% in 1990 and 81% in the 2000s. In 2012, 136 million tons of world fish production accounted for 86% utilization for direct human consumption, while the remaining 14% were for nonfood purposes including reduction to fishmeal and fish oil, ornamental purposes, pharmaceutical uses, or animal feed.

Live forms of fish, fresh or chilled, are usually preferred and comprise the higher-priced products in many markets; the proportion of live fish marketed for edible purposes in 2012 reached about 46% (63 million tons). Other forms of fish, such as dried, salted and smoked, accounted for 12% (16 million tons), 13% (17 million tons) were prepared and preserved forms, and the remaining 29% (40 million tons) were frozen products (FAO 2014).

From 1976 to 2012, the trade of fish and fishery products increased by about 8.3% per year in nominal terms and 4.1% in real terms. The international financial crisis which began in 2008 and its subsequent economic contraction in several economies created a pressure in the market, leading to reduced demand from key markets. Pressure to reduce prices in some fishery products, especially in farmed species, has also been noticed. The combination of these factors can explain the stall at international trade from the 2011 peak at US$ 129.8 billion (which was an 17% increase over 2010), to US$ 129.2 billion in 2012. In 2012, 90% of trade in fish and fishery products consisted of processed products in terms of quantity (live weight equivalent[1] excluding live and fresh whole fish), and only the remaining 10% were of live fish (fresh and chilled). Nevertheless, it is clear that in the last decades there was an improvement in logistics and a growing demand for unprocessed fish, since it doubled the 5% value of 1976.

---

[1]Mass or weight, when removed from the water.

Preliminary statistics indicate an increase in world per capita fish consumption of 9.9 kg in 1960 to 19.2 kg in 2012. The presence of fish in a healthy diet is so important that, in 2010, its consumption was responsible for 16.7% of the global population's intake of animal protein and 6.5% of all protein consumed. Fish proteins therefore represent an essential nutritional component in countries with high population density, where the total level of protein intake may be low (FAO 2014).

### 1.2.3 Fish production

The last five decades have seen a growing world fish production, increasing the supply of fish food at an average annual rate of 3.2%, outpacing population growth worldwide at 1.6%. World food fish aquaculture production grew at an average annual rate of 6.2% between 2000 and 2012, more than doubling its value from 32.4 million tons in 2000 to 66.6 million tons in 2012.

The global capture fishery production, in both marine waters and inland waters, has remained stable over the last 5 years, reaching the second highest global capture fishery production of 93.7 million tons in 2011 and 91.7 million tons in 2012. Comparing capture production with aquaculture production, capture production stabilized in the 1990s while aquaculture production continued to grow. In fact, the global aquaculture production (for human consumption and other uses) reached a new high of 90.4 million tons (live weight equivalent) in 2012, and FAO estimates that aquaculture production for human consumption will increase 5.4% to 70.2 million tons in 2013. In fact, in Asia the aquaculture production has surpassed the capture production since 2008 (FAO 2014).

### 1.2.4 Fish as a source of income

Employment in the fisheries sector grew at a faster rate than employment in the traditional agriculture sector during the period between 1990 and 2012. Approximately 58.3 million people were estimated to be involved in the primary sector of capture fisheries and aquaculture in 2012, representing about 4.4% of the 1.3 billion people economically active in the agriculture sector; these figures are well above the 2.7 and 3.8% registered in 1990 and 2000, respectively.

The growing importance of aquaculture in the fisheries sector can also be observed in the employment. In the last decade, the average annual percentage growth rate of fish farmers was 3.7% and 4.1% during 2000–2005 and 2005–2010, respectively, while in capture fishers it was only of 1.2% and 1.5% in the same periods. As well as fish farmers and catchers, this sector also provides jobs in the secondary sector; overall, the fisheries sector provides employment to 10–12% of world's population (FAO 2012).

### 1.2.5 World fleet

The total number of fishing vessels in the world was estimated at 4.72 million in 2012. The distribution of the global fleet by region designates Asia as the biggest player in the world with 3.23 million of vessels (68%). In second place comes Africa with 16% (mostly due to inland, nonmotorized small vessels), followed by

Latin America and the Caribbean, North America, and Europe, with 8%, 2.5%, and 2.3%, respectively. Although the inland fleet is significant, representing 32% of the world fleet in 2012, its proportion relative to marine vessels operating in inland waters varied substantially by region, being the highest in Africa (64%), followed by Asia (30%), and Latin America and the Caribbean (18%) (FAO 2014).

### 1.2.6 The status of fishery resources

Since 1996 when marine fisheries reached a maximum production of 86.4 million tons, the total capture of fish has been steadily declining, dropping to 79.7 million tons in 2012.

In 2011, the stocks of fish captured within biologically sustainable levels were 71.2%; this means that 28.8% was fished at levels where the abundance is lower than the level that can produce the maximum sustainable yield, being almost 3 times higher than in 1974 when it was about 10%.

There is in fact no perspective that in the near future this scenario will change; it is actually expected that this trend in reducing marine fisheries resources will continue since, in 2011, only 10% of such resources were underfished (FAO 2014).

### 1.2.7 Unveiling the future

According to the World Bank, five years after the global financial crisis, the world economy shows signs of regression in 2014. However, thanks to general signs of growth, preliminary estimates for 2013 indicate a further increase of fish trade and fishery products. It is expected that the long-term slow but steady economic recovery will continue to show a positive trend for fish trade in the future, despite the instability in stocks during the last couple of years (FAO 2014).

Aquaculture is at a historical turning point since it is estimated to surpass capture as the main source of fish for human consumption in 2015 (OECD & FAO 2013). Nevertheless, it will face some challenges that are responsible for the estimated reduction in its growth from the 5.6% per annum in the last decade to 2.5% per annum until 2023 (OECD & FAO 2014). The main restrictions are:

- *regulation and competition for space*: especially in coastal areas, competition with other users will increase and licensing may become more difficult to obtain;
- *investment in research*: an increasing investment in research will be necessary to increase productivity, find substitutes for fishmeal and fish oil, increase animal health and avoid diseases, and reduce environmental impact;
- *production costs*: expected increase in fishmeal, fish oil, and energy prices; and
- *climate change*: offshore aquaculture will be affected by the rising temperatures of the oceans.

The reduction in the aquaculture growth trend, together with an almost stable capture production, means that only an increase of 2.5% is estimated for the period of 2014–2023 (OECD & FAO 2014). This will lead to a rise in prices (Figure 1.5) since demand will increase at the same time not only by world population growth, but also by *per capita* consumption increase (from the 19.2 kg in 2011–2013 to 20.9 kg in 2023). Demand pressure has a tendency to be reduced

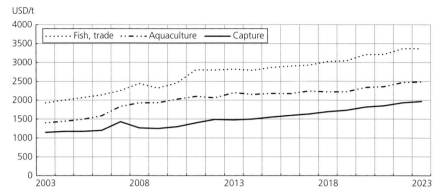

**Figure 1.5** World fish price development in nominal terms between 2003 and 2023. Source: OECD.

in the future because, even with the greater awareness of the general population towards the importance and great benefits of fish protein on human diet, the higher price of fish will lead to its substitution by other cheaper sources. Moreover, although the population rate is growing, that growth will tend to become slower (in the last 10 years the growth rate has fallen from 1.24% to 1.18%), which will also contribute to the deceleration of the demand pressure (UN 2015).

The world total fisheries production is projected to reach 181 million tons in 2022. In 2013, exports reached a new record of over US$ 136 billion, larger than the previous year by more than 5%. Despite the decline in the average growth rate projected for the 2013–2022 period, fish products will still be one of the most highly traded goods in the word with exports reaching 36% of total fishery production in 2022 (FAO 2014).

## 1.3 Hazard assessment in seafood

Hazards can be classified as one of three distinct categories: biological, chemical, and physical hazards. In the particular case of fish, biological hazards include pathogenic bacteria, viruses, and parasites. Chemical hazards refer to pesticides, polychlorinated biphenyls (PCBs), heavy metals, veterinary medicines, dioxins, polycyclic aromatic hydrocarbons, marine biotoxins and biogenic amines (including histamine; note that these are sometimes classified as biological hazards as in CAC 2003), allergens, chemical from materials in contact with food-stuffs, waste products, and cleaning compounds. Finally, physical hazards such as hooks, fishing nets, plastic, pieces of glass, wood, personal ornaments, and constituent pieces of equipment are commonly identified. These are all described in more detail in Sections 1.3.1–1.3.3 below.

In order to prevent foodborne illnesses caused by ingestion of food contami-nated by pathogenic bacteria, both the WHO and the FDA advise cooking foods until they reach temperatures above 70°C (140°F) at the thermal center of the product (WHO 2006; FDA 2014).

## 1.3.1 Biological hazards
### Pathogenic bacteria

Pathogenic bacteria are the main agents responsible for cases of infections/food poisoning. Food infections are characterized by ingestion of viable bacteria, which multiply inside the human body. On the other hand, food poisoning is caused by the ingestion of food previously contaminated with toxins produced by pathogenic bacteria, which are present in high numbers in food (Huss *et al.* 2004).

According to Huss *et al.* (2004), pathogenic bacteria associated with diseases caused by the ingestion of fish can be divided into three major groups:

- indigenous pathogenic bacteria of the aquatic environment: *Clostridium botulinum* (nonproteolytic type B, E, and F), *Vibrio* (*V. cholerae, V. parahaemolyticus*, and *V. vulnificus*), and *Plesiomonas shigelloides*;
- indigenous pathogenic bacteria of the general environment: *Listeria monocytogenes, Clostridium botulinum* (proteolytic type A and B), *Clostridium perfringens*, and *Bacillus spp.*; and
- indigenous pathogenic bacteria of humans and animals: *Salmonella spp., Shigella spp., Escherichia coli, Campylobacter jejuni, Yersinia enterocolitica*, and *Staphylococcus aureus*.

Although several factors can influence the growth rate of microorganisms, temperature is undoubtedly one of the most relevant. For that reason, the minimal growth temperatures of pathogenic bacteria associated with diseases caused by contaminated fish ingestion are listed in Table 1.2. It is clear from this set of data that there is no consensus regarding the minimum temperature needed for the growth of certain pathogenic bacteria; the lowest temperature should therefore be considered for the hazard analysis.

### Escherichia coli

The genus *Escherichia* is a group of Gram-negative bacteria which are facultative anaerobics, mesophilic, rod-shaped organisms with an optimum growth temperature of 37°C (98.6°F) (Delgado 2006).

*Escherichia coli* belong to the family *Enterobacteriaceae*, and are one of the most predominant species in the intestinal tract of humans and warm-blooded animals as part of natural intestinal flora. In addition to being harmless to the host, this species provides multiple benefits such as prevention of gut colonization by various pathogenic bacteria (FDA 2012a). However, there is a small group of bacteria within the *E.coli* species that are pathogenic for humans. According to Huss *et al.* (2004), the main groups of *E. coli* responsible for cases of food poisoning are:

- enterotoxigenic *E. coli* (ETEC);
- enteropathogenic *E. coli* (EPEC);
- enterohemorrhagic *E. coli* (EHEC);

**Table 1.2** Minimal growth temperatures of pathogenic bacteria associated with seafood

| Microorganism | Temperature (°C/°F) | | |
|---|---|---|---|
| | FDA (2011) | Jay *et al.* (2005) | Huss *et al.* (2004) |
| *Clostridium botulinum* (proteolytic, types A and B) | 10/50 | – | 10/50 |
| *Clostridium botulinum* (nonproteolytic, types B, E and F) | 3.3/37.4 | — | 3.3/37.4 |
| *Clostridium perfringens* | 10/50 | — | – |
| *Vibrio cholerae* | 10/50 | — | 10/50 |
| *V. parahaemolyticus* | 5/41 | 5/41 | 5/41 |
| *V. vulnificus* | 8/46.4 | – | 8/46.4 |
| *Vibrio spp.* | – | – 5/23 | – |
| *Plesiomonas shigelloides* | – | – | 8/46.4 |
| *Listeria monocytogenes* | – 0.4/31.28 | 1/33.8 | 0–2/32–35.6 |
| *Salmonella spp.* | 5.2/41.36 | 7/44.6 | 5/41 |
| *Shigella spp.* | 6.1/42.98 | – | 6 |
| *Escherichia coli* | 6.5/43.7 | – | 7/44.6 |
| *Staphylococcus aureus* | 7/44.6 | 6.7/44.06 | 7/44.6 |
| (toxin production) | 10/50 | – | 10/50 |
| *Yersinia enterocolitica* | – 1.3/29.66 | – 2/28.4 | – 1.3/29.66 |
| *Bacillus cereus* | 4/39.2 | 7/44.6 | – |

- enteroinvasive *E. coli* (EIEC);
- enteroaggregative *E. coli* (EAEC); and
- diffusely adherent *E. coli* (DAEC).

Of those groups, ETEC, EPEC, EIEC, and EHEC *E. coli* can be transmitted to humans through contaminated food and water. EHEC *E. coli*, more specifically *E. coli* serotype O157:H7, is more dangerous due to its ability to produce verotoxin (FDA 2012a). The most common symptoms associated with this serotype are hemorrhagic colitis and hemolytic uremic syndrome (FDA 2011).

When microbiological analyses of food products are performed, *E. coli* is commonly used as an indicator of fecal contamination with the possible presence of enteric pathogens.

### Staphylococcus aureus

Bacteria of the genus *Staphylococcus* are generally Gram-positive, are of spherical shape (coconut), are immobile, and group themselves in the form of aggregates. *Staphylococcus aureus* belong to the family *Micrococcaceae* and have an optimum growth temperature between 35°C (95°F) and 37°C (98.6°F) (Delgado 2006). *S. aureus* are often found in soil, water, air, skin, and mucous glands of humans and warm-blooded animals, as well as in all surfaces that come into contact with them (Huss *et al.* 2004). Many *Staphylococcus* species, including coagulase-positive and coagulase-negative strains, have the ability to produce extremely heat-stable enterotoxins (FDA 2012a); however, according to Huss *et al.* (2004), this only occurs when bacterial concentrations exceed $10^6$ CFU/g. The main symptoms of poisoning caused by *S. aureus* are abdominal pain, diarrhea, vomiting, nausea, weakness and, in some cases, death (FDA 2011).

### Salmonella spp.

Bacteria belonging to the genus *Salmonella* are mesophilic, Gram-negative, movable, rod-shaped, non-spore-forming organisms that have their optimum growth temperature at about 37°C (98.6°F) (Delgado 2006). *Salmonella* belong to the family *Enterobacteriaceae* and can be found in the intestinal tract of humans and animals in general. This genus can be divided into two species that are pathogenic to humans: *Salmonella enterica* and *Salmonella bongori*. The first species constitutes a greater public health concern and is subdivided into six subspecies (FDA 2012a):

- *S. enterica* subsp. *enterica* (I);
- *S. enterica* subsp. *salamae* (II);
- *S. enterica* subsp. *arizonae* (IIIa);
- *S. enterica* subsp. *diarizonae* (IIIb);
- *S. enterica* subsp. *houtenae* (IV); and
- *S. enterica* subsp. *indica* (VI).

These can be further subdivided into more than 2500 known serotypes, such as *S. enterica* sp. *enterica* serotype Typhimurium (*S.* typhimurium). Depending on the serotype, *Salmonella* can cause two types of disease – nontyphoid and typhoid Salmonellosis – with the latter leading to a higher mortality rate (FDA 2012a). The main symptoms are high typhoid fever, headache, diarrhea or constipation and, in some cases, may lead to death (FDA 2011).

### Listeria monocytogenes

*Listeria monocytogenes* are Gram-positive, facultative anaerobic bacteria, which occur in rod-shaped mobile cells (bacillus) due to the presence of flagella. *L. monocytogenes* belong to the family *Listeriaceae* and are ubiquitously distributed in nature (Huss *et al.* 2004; FDA 2012a). They can grow at 37°C (98.6°F) and are halotolerant and psychrotolerant (Huss *et al.* 2004). Seven species of *Listeria* are known; however, only *L. monocytogenes* is pathogenic to humans, which is the main cause of death from foodborne diseases. This species can be divided into 13 serotypes (1/2a, 1/2b, 1/2c, 3a, 3b, 3c, 4a, 4ab, 4b, 4c, 4d, 4e, and 7), with the majority of food infections associated with serotypes 1/2a, 1/2b, and 4b (FDA 2012a).

The main symptoms of infection by *L. monocytogenes* are flu syndrome, meningitis, septicemia, encephalitis, spontaneous abortion, and fetal death in the uterus. Pregnant women, newborns, and people with weak immune system are more susceptible to this type of bacteria (FDA 2011).

### Vibrio parahaemolyticus

Bacteria of the genus *Vibrio* belong to the family *Vibrionaceae*, are generally found in marine environments and/or estuaries, and the majority of them require the presence of sodium chloride to grow (FDA 2012a). *V. parahaemolyticus* are

Gram-negative, rod-shaped (bacilli), facultative anaerobic, mesophilic bacteria. Not all pathogenic strains are characterized by the production of hemolysins (e.g., Thermostable Direct Hemolysin (TDH) and/or Thermo Stable Related Hemolysin (TRH)), responsible for the lysis of red blood cells (FDA 2012a). The most common symptoms are diarrhea, abdominal cramps, fever, headache, vomiting, chills, nausea, and primary septicemia (FDA 2011).

Of the 34 known *Vibrio* species, 13 are pathogenic to humans and *V. parahaemolyticus, V. cholerae,* and *V. vilnificus* are the main species associated with cases of foodborne illnesses caused by eating fish. It should be noted that all cases related to the presence of *V. parahaemolyticus* were caused by the consumption of fish (Huss *et al.* 2004).

## Viruses

Viruses are small microorganisms (25–70 nm) that can only reproduce within a host; the number of viral particles in the food therefore remains constant. It is estimated that in the aquatic ecosystem there are approximately 10 thousand million viruses per liter of water, but none are pathogenic to humans.

Viruses associated with foodborne illness are called enteric viruses (originated in the gastrointestinal tract of humans) and their presence in fish is mainly due to poor hygienic conditions of food handlers or poor water quality (pollution of the aquatic environment with waste water) (Huss *et al.* 2004).

According to the FDA (2011), the main viruses associated with diseases caused by fish consumption are:

- *Norwalk* virus, which causes symptoms such as diarrhea, nausea, vomiting, abdominal cramps, headache, and a low-fever painful body; and
- Hepatitis A, for which symptoms include nausea, malaise, abdominal discomfort, anorexia, and jaundice.

When compared to pathogenic bacteria, these viruses exhibit a higher resistance to low temperatures; they are stable at refrigeration temperatures, and freezing temperatures only lead to a small increase in the rate of inactivation. Consumption of raw or undercooked bivalve mollusks, especially oysters, is the main source of viral contamination associated with fish (FAO & WHO 2010).

## Parasites

Parasites are often found in seafood (with the exception of bivalve mollusks) and, according to the FDA (2011), these can be divided into three groups as follows.

- Nematodes (including *Anisakis spp.* and *Pseudoterranova spp.*): the *Anisakis simplex* and *Pseudoterranova decipiens* species are commonly found in fish. These parasites cause nausea, vomiting, diarrhea, severe abdominal pain, and may also penetrate the gut.
- Cestodes (e.g., *Diphyllobothrium spp.*): these cause symptoms such as abdominal pain, weight loss, and anemia.
- Trematodes (*Heterophyes spp.* and *Nanophyetes salmincola*): these cause abdominal pain and diarrhea.

Parasites are transmitted to humans through the consumption of raw or undercooked fish. Usually, heat treatment (60°C/140°F for 1 minute) is effective in the elimination of parasites present in food. Freezing can also be used for parasite elimination although its effectiveness depends on several factors, such as (FDA 2011):

- type of fish;
- type of parasite;
- temperature of freezing;
- time needed to freeze the fish; or
- the period of time during which the fish remains frozen.

According to Regulation (EU) No. 1276/2011, in order to eliminate the parasitic tapeworms and roundworms the fish should be submitted to a freezing treatment which consists of lowering the temperature in all parts of the fish to –20°C (–4°F) for at least 24 hours, or to –35°C (–31°F) for at least 15 hours. On the other hand, the FDA states that freezing and storing at an ambient temperature of –20°C (–4°F) or below for a total of 7 days, freezing at an ambient temperature of –35°C (–31°F) or below until solid and storing at an ambient temperature of –35°C (–31°F) or below for 15 hours, or freezing at an ambient temperature of –35°C (–31°F) or below until solid and storing at an ambient temperature of –20°C (–4°F) or below for 24 hours are sufficient to kill parasites (FDA 2011).

### 1.3.2 Chemical hazards

According to Huss *et al.* (2004), chemical compounds may or may not be toxic to human health, depending on their concentration. The most severe effects, such as neurological damage, birth defects, and/or oncological diseases, occur when the organism is exposed to low concentrations for a long period of time.

Environmental contamination with chemical compounds such as pesticides is a direct result of human activity. These compounds are present in low levels in lakes, rivers, seas, or oceans and accumulate particularly in fish by biomagnification (the concentration increases along the food chain from one trophic level to the next) or by bioaccumulation (concentration increases throughout the fish life by exposure to contaminated environments). The risk of contamination is relatively low in wild fish captured offshore, but increases in coastal locations or in fish originating from aquaculture as a result of human activity.

### Pesticides

Pesticides are chemicals used in horticultural and agricultural soils to combat pests (e.g., insects) and pathogenic microorganisms and in aquaculture systems to eliminate weeds, algae, and invertebrates. These substances accumulate in fish and can be transmitted to humans through its ingestion (FDA 2011; WHO 2010d).

Pesticides can affect the nervous, cardiovascular, reproductive, and endocrine systems, the gastrointestinal tract, liver, and kidneys, and some may be carcinogenic (FDA 2011).

## Polychlorinated biphenyls (PCBs)

Polychlorinated biphenyls are fat-soluble environmental contaminants which are able to produce persistent organic pollutants due to their chemical and physical properties, such as nonflammability, stability, high boiling point, low heat conductivity, and a high dielectric constant. These products are used as flame-retardants in paints, heat transfer systems, hydraulic fluids, lubricants, and dielectric fluids in capacitors and transformers, and are stored in adipose tissues of the body since their absorption occurs through the respiratory system, gastrointestinal tract, or skin. In order to reduce the levels of these products, the skin and trimming fat of fish exposed to them must be removed before cooking. Certain cooking procedures, such as using grills, can also minimize PCB exposure. This exposition causes neurobehavioral alterations, including abnormal reflexes, motor immaturity, slowed mental development, alteration in memory function, and decreased immunocompetence. In order to avoid the risk of PCB contamination, two professional groups – Physicians for Social Responsibility and The Association of Reproductive Health Professionals (ARHP) – recommend limiting the intake of fatty fish to one or two times per month (Sidhu 2003; Dovydaitis 2008).

## Heavy metals

According to Huss *et al.* (2004), of the several heavy metals in the environment those exhibiting a greater concern to human health include arsenic, cadmium, lead, and mercury, as discussed in the following sections.

## Arsenic

Arsenic is naturally distributed in the Earth's surface, usually as arsenic sulfides or arsenates and metal arsenides. This pollutant is released into the atmosphere as a result of natural activities (e.g., volcanic activity) and anthropogenic activities (e.g., burning of fossil fuels). Processes involving high temperatures mainly generate arsenic trioxide, an inorganic compound adsorbed by particles dispersed in the air. These particles are carried by wind and may be deposited in terrestrial and aquatic ecosystems (WHO 2010a).

Of the various arsenic compounds present in the environment, organic compounds (e.g., arsenobetaine) have lower toxicity to living organisms. In fish there is a predominance of arsenobetaine. However, there are no current studies that prove the absence of inorganic compounds in the foodstuff as well as the possible conversion of arsenobetaine into toxic compounds by the human body (Ahmed 1991).

## Cadmium

Cadmium is a metal present in low concentrations in the environment and results from various human activities, such as the burning of fossil fuels and incineration of municipal waste, as well as natural activities (e.g., volcanic activity, weathering, runoff). The emergence of the industrial era increased the load of this and other metals used in industrial processes in the environment, resulting in higher levels of pollution from improper disposal of wastes after use in industry (James 2013).

This compound can be atmospherically transported over long distances and/or deposited in the soil and/or groundwater, accumulating rapidly in living organisms including mollusks and crustaceans. In fact, cadmium accumulates in different levels in mammalian kidneys, hepatopancreas and muscles of some mollusks, large swordfish, marlin, and tuna (James 2013). When accumulated in the kidneys, this metal can cause renal tubular dysfunction. Apart from the kidneys, this metal has toxic effects in the respiratory system and skeleton (James 2013). It is also considered carcinogenic, causing throat, kidney, and prostate cancer (WHO 2010b).

### Lead

Lead is a heavy metal found in low concentrations on the Earth's surface mainly in the form of lead sulfide. This compound has a similar behavior to cadmium in terms of transport and accumulation.

This metal is responsible for neurological disorders, anemia, headache, irritability, lethargy, convulsions, muscle weakness, ataxia, tremors, and paralysis. In men, it affects the quality and quantity of sperm, whereas in pregnant women it can cause miscarriage, fetal death, pre-term birth (low birth weight), and birth defects. Lead is also considered carcinogenic (WHO 2010e).

### Mercury

Mercury is a reactive heavy metal emitted from natural sources (e.g., volcanoes, forest fires) and human sources (e.g., coal-fired electric power plants, gold mining, industrial boilers, chlorine production, waste incineration). It is present in nature in various forms, including elemental, organic compounds (e.g., methyl mercury and ethyl mercury), and inorganic compounds (e.g., mercuric chloride). Elemental mercury is present in a liquid state at room temperature; however, due to its high volatility, it easily passes to the gas phase. This gas remains in the atmosphere and can be deposited in the aquatic environment (e.g., rivers, lakes), where it is transformed into methyl mercury.

Mercury in the Earth's crust is mostly in the inorganic form, but once in the aquatic environment it can change to methyl mercury. This form is soluble, mobile, and quickly enters the aquatic food chain, accumulating in a greater extent in biological tissues than inorganic forms of mercury, and constitutes over 90% of the total mercury detected in fish (James 2013). Methyl mercury is absorbed by phytoplankton and accumulates in fish, primarily in the predatory species that have a longer lifetime, such as shark and swordfish. Methyl mercury follows the food chain passing through bacteria, plankton, macro invertebrates, herbivorous fish, piscivorous fish, and finally humans (Dovydaitis 2008). Worldwide emissions, natural sources, and the environmental half-life of the compound preclude short-term reductions, although it may be possible to reduce global average methyl mercury concentrations in fish in the long term (James 2013).

Increased levels of fish consumption during pregnancy and lactation have been linked to a significant reduction in incidences of pre-term labor and delivery, the risk of intrauterine growth restriction, and pregnancy-induced hypertension, and are also associated with increases in a child's intelligence quotient (IQ) (Sidhu 2003). Nevertheless, both elemental mercury and methyl mercury exhibit high toxicity to the central and peripheral nervous systems because mercuric ions concentrate in the brain. All forms of mercury can cause neurological and behavioral disorders, among which memory loss and cognitive and motor dysfunction stand out, particularly in children. It can also cause seizure disorders, blindness and choreoathetosis (Dovydaitis 2008). When pregnant women ingest food contaminated with methyl mercury, this metal can lead to the development of neurological disorders in the fetus including mental retardation, seizures, vision and hearing loss, developmental delay, speech disturbances, and memory loss (WHO 2007).

## Veterinary medicines

In recent years there has been a significant increase in the use of veterinary drugs in the aquaculture system. These accumulate in the edible parts of fish and are transmitted to humans through the ingestion of food (Huss *et al.* 2004). In aquaculture systems, the use of veterinary medicines has the following main objectives (Huss *et al.* 2004; FDA 2011):

- prevention and treatment of diseases in fish caused by a wide variety of pathogenic bacteria, such as *Aeromonas hydrophila*, *A. salmonicida*, *Edwardsiella tarda*, *Pasteurella piscicida*, *Vibrio anguillarum*, *V. salmonicida*, and *Yersinia ruckeri*;
- control of the growth and reproduction of fish; and
- control of parasites.

The use of pharmacologically active substances in veterinary medicines is commonly controlled and regulated by authorities. Even when their use is allowed, there are maximum levels of its presence in various foodstuffs that must be complied with. In the European Union for example, pharmacologically active substances are regulated by Regulation (EU) No. 37/2010 (EC 2010).

Veterinary drugs may cause allergies and intestinal flora modifications, and reduce the efficacy of antibiotics in the treatment of infections due to acquired resistance (Huss *et al.* 2004).

## Dioxins

Dioxins are toxic chemical compounds formed during the combustion of organic compounds in the presence of chlorine. The term dioxin includes 75 polychlorinated dibenzo-para-dioxins (PCDDs), 135 polychlorinated dibenzofurans (PCDFs), and 12 polychlorinated biphenyls (PCBs) in the form of dioxins, all presenting a similar mechanism of action and toxicological properties (EFSA 2010; WHO 2010c).

Using several routes such as air, soil, and water, dioxins and dioxin-like PCBs cause environmental contamination to reach the food chain (Sidhu 2003). The main source of contamination of the environment and foodstuffs results from dioxin release from industrial plants, where food consumption is the main form of exposure of humans. According to Sidhu (2003) about 95–98% of human exposure comes from the food supply, mostly from products of animal origin (e.g., dairy products, seafood, meat). These compounds remain for long periods of time in the atmosphere, soil, and aquatic ecosystems, therefore considered persistent organic compounds (POPs), and accumulate in food with high fat content (such as fish) due to their low solubility in water (WHO 2010c).

Dioxins can affect the developing nervous system, thyroid, and reproductive functions (WHO 2010c).

## Polycyclic aromatic hydrocarbons

Polycyclic aromatic hydrocarbons (PAHs) are compounds formed during incomplete combustion, pyrolysis of organic matter, or even as a result of thermal processing. PAHs have a high persistence in the environment and can contaminate aquatic ecosystems and accumulate in living organisms, particularly in fish. PAHs are considered carcinogenic and genotoxic compounds (EFSA 2008).

Benzo(a)pyrene (BaP) is the most-studied aromatic hydrocarbon and, when limits are defined, it can be used as a marker regarding the occurrence of carcinogenic PAHs in food.

## Marine biotoxins

Marine biotoxins are toxic substances produced by certain marine algae (phytoplankton), indigenous bacteria from the aquatic environment such as *Vibrio alginolyticus*, *Bacillus* spp., and *Pseudomonas* spp., and species of fish. In the first case, it is estimated that of the 4000 known species of seaweeds, only 70–80 species produce biotoxins. These toxins accumulate mainly in bivalve mollusks by filtration or may accumulate in certain species of fish when they ingest toxic algae or other previously contaminated species. Thermal processing, even at high temperatures such as frying, is usually not sufficient to eliminate biotoxins (Huss *et al.* 2004).

A description of the most common marine biotoxins in bivalve mollusks is provided in the following. The maximum legal limits for their presence can be found, for example, in Regulation (EC) No. 853/2004 and in the FDA's Fish and Fishery Products Hazards and Controls Guidance (EC & EP 2004; FDA 2011).

- Saxitoxins cause poisoning of the type Paralytic Shellfish Poison (PSP). Main symptoms of PSP are drowsiness, tingling, numbness, burning sensation on the lips and tongue extending to the face and hands, and lack of coordination of the hands, legs, and chest. In most severe cases, PSP can lead to respiratory arrest and death (FDA 2011).
- Domoic acid causes poisoning of the type Amnesic Shellfish Poison (ASP). This type of poisoning is responsible for symptoms such as nausea, vomiting, diarrhea, abdominal cramps, headache, and some neurological effects (e.g., loss of short-term memory, disorientation, dizziness, confusion). In more severe cases, seizures, coma, and death may occur (FDA 2011).

- Okadaic acid and dinophysistoxins cause poisoning of the type Diarrhoeic Shellfish Poisoning (DSP). This disorder is essentially responsible for gastrointestinal disorders, abdominal pain, nausea, vomiting, diarrhea, headache, and fever (FDA 2011).
- Azaspiracids cause poisoning of the type Azaspiracid Shellfish Poisoning (AZP). The main symptoms of AZP are abdominal pain, nausea, vomiting, and diarrhea (FDA 2011).

Apart from the marine biotoxins described above, the FDA also mentions other biotoxins including ciguatoxin, brevetoxin, and tetrodotoxin (FDA 2011). In the case of the EU, to prevent and control the occurrence of disease incidences related to the ingestion of shellfish contaminated with marine biotoxins, Regulation (EC) No. 853/2004 states that fish farmers can only harvest in production areas with locations and boundaries fixed by competent authorities (EC & EP 2004). Given the level of fecal contamination of bivalve mollusks, competent authorities classify the production areas as one of three categories (A, B, and C). Mollusks from Class A production areas can be placed directly into the market, whereas mollusks from Class B and Class C areas may only be marketed after decontamination in purification centers or relaying areas to reduce and/or eliminate contaminants (EC & EP 2004). According to the Regulation (EC) No. 854/2004, production and relaying areas should be regularly monitored by authorities that monitor, among others, the microbiological quality of live bivalve mollusks, the presence of toxin-producing plankton in the waters, and the presence of biotoxins in bivalve mollusks.

## Histamine

Histamine is a biogenic amine called scombrotoxin produced in the *post-mortem* period of fish. This amine results from the decarboxylation of histidine (an amino acid present in high and variable amounts in fish) into histamine by the enzyme histidine decarboxylase. Once present in fish, this enzyme is responsible for the ongoing formation of histamine even at refrigeration temperatures. Its inactivation occurs at freezing temperatures, but it can be reactivated rapidly after thawing. It should be noted that histamine is not removed by thermal processing such as cooking (FDA 2011).

Histidine decarboxylase is produced by bacteria found in saltwater environments (e.g., *Vibrio* spp.) and can therefore be present on the surface or inside the fish body (gills, liver, and/or intestine) (FDA 2011). This enzyme may also be produced by bacteria that contaminate fish after being caught, such as *Clostridium perfringens* (Huss *et al.* 2004). The bacteria responsible for the production of the histidine decarboxylase are able to multiply over a wide temperature range, although a more rapid formation of histamine is achieved at higher temperatures (21.1°C/69.98°F or higher) than at moderate temperatures (7.2°C/44.96°F). In order to inactivate some of these bacteria, fish should be maintained at freezing temperatures. Bacteria and histidine decarboxylase enzymes are efficiently inactivated by thermal processing such as cooking (FDA 2011).

High levels of histamine are commonly found in fish species of the families *Scombridae, Clupeidae, Engraulidae, Coryfenidae, Pomatomidae*, and *Scombresosidae* (EC 2007). The consumption of fish contaminated with histamine causes tingling or burning around the mouth or throat, skin rash or hives, low blood pressure, headache, dizziness, itchy skin, nausea, vomiting, diarrhea, asthma, heart palpitations, and breathing problems (FDA 2011).

## Allergens

There are certain components present in food which are capable of inducing an immune response. In the United States of America, the Food Allergen Labeling and Consumer Protection Act (FALCPA) (Public Law 107-282) that took effect on 1 January 2006 lists eight foods that are considered 'major food allergens' and should therefore be declared on food labels. The 'Big Eight' are: milk, eggs, fish, crustacean shellfish, tree nuts, peanuts, wheat, and soybeans (Helm & Burks 2009). The *Codex Alimentarius* Commission issued general standards for the labeling of packaged food and listed the same eight substance categories (CAC 2010).

Due to the fact that the prevalence of food allergy varies in different countries, some governments have added additional foods to the list of ingredients that must be declared on food labels (Helm & Burks 2009). For example, the European Union has a list of 14 allergenic substances, which can be found in Regulation (EU) No. 1169/2011. The list includes (EC & EP 2011):

- cereals containing gluten and products based on these cereals;
- crustaceans and crustacean-based products;
- eggs and egg-based products;
- fish and fish-based products;
- peanuts and peanut-based products;
- soybeans and soybean-based products;
- milk and milk-based products (including lactose);
- nuts and nut-based products;
- celery and celery-based products;
- mustard and mustard-based products;
- sesame seeds and sesame-seed-based products;
- sulfur dioxide and sulfites in concentrations higher than 10 mg/kg or 10 mg/L expressed as total $SO_2$;
- lupin and lupin-based products;
- mollusks and mollusk-based products.

Allergens cause tingling in the mouth, swelling of the tongue and throat, difficulty in breathing, hives, vomiting, abdominal cramps, diarrhea, low blood pressure, loss of consciousness and, in the most severe cases, death (FDA 2011).

## Food additives

In accordance with the General Standard for Food Additives (FAO & WHO 2014), a food additive can be defined as:

'any substance not normally consumed as a food by itself and not normally used as a typical ingredient of the food, whether or not it has a nutritive value, the intentional addition of which to food for a technological (including organoleptic) purpose in the manufacture, processing, preparation, treatment, packing, packaging, transport or holding of such food results, or may be reasonably expected to result (directly or indirectly), in its byproducts becoming a component of or otherwise affecting the characteristics of such foods'.[2]

Food additives can be used as preservatives, antioxidants, acidity regulators, and acidifiers. Those that represent minimum toxicological concerns, such as calcium carbonate or lactic acid, can be added to the vast majority of processed food, while the use of those which pose a greater risk for the health of consumers should be limited (EC 2011).

In the General Standard for Food Additives, a list of food additives that can be used in foods, which product category they may be used in, and their maximum concentration in the product is provided. The Australian New Zealand Food Standards Code (Standard 1.3.1 Food Additives, Volume 2) also has a list of products in which food additives can be used and which food additives are allowed to be used in each of the products. In the European Union the safe food additives list is provided in Regulation No. 1333/2008 and respective amendments (in particular Regulation (EU) No. 1129/2011). In this regulation a list of additives that are allowed in fish and fishery products can be found, for example sulfur dioxide (sulfites in crustaceans are used as a way to prevent the formation of 'black spot') (FDA 2011). Regular consultation of these lists is recommended as they are frequently updated.

When an overdose of both authorized and unauthorized food additives occurs, foods become unsafe for human consumption. Additives can cause hypersensitivity reactions or food intolerance, tingling sensation in the mouth, swelling of the tongue and throat, breathing problems, hives, vomiting, abdominal cramps, and diarrhea. Some additives may also be carcinogenic (FDA 2011).

### Plastic materials and objects in contact with foodstuffs

Chemical compounds from plastic materials and objects that can migrate to food are considered chemical hazards. In this sense, it is important not to use substances that can constitute a hazard to human health as a constituent of plastic materials that are in contact with food products.

There are several published standards that identify those substances. For example, in 2008 the Chinese Ministry of Health (MOH) published the GB 9685-2008 *Hygienic Standard for Use of Additives in Food Containers and Packaging Materials*, which contains over 1500 substances approved to be used in China.

The European Union has also published a list of authorized substances in the manufacture of plastic layers in plastic materials and articles in Regulation (EU) No. 10/2011 and respective amendments (in particular, Regulation (EU) 2015/174).

---

[2] According to the standard, this definition does not include contaminants or substances added to food for maintaining or improving nutritional qualities.

According to this Regulation, packaging materials contain plastic additives and monomers, especially vinyl chloride (a substance considered carcinogenic that, by polymerization, yields polyvinyl chloride). It should also be taken into account that the plastic can be printed, coated or held together by adhesives.

### Waste products from chemical cleaning agents and equipment
Among all products for cleaning and disinfection, biocidal products used in the fight against pests stand out. Only active substances authorized for the purpose should be present in their composition (in the case of the EU, these substances are regulated by (EC) No. 528/2012).

In case of errors in the dosage of biocidal products or poor elimination, surfaces that come into contact with food can become contaminated with residues of these products, constituting a source of food contamination.

Chemical components of some equipment, such as lubricating oils, are also considered chemical hazards. The presence or prohibition of some constituents is currently not internationally consensual, but it is possible to find a reference in the published FDA Code of Federal Regulations Title 21, Sec. 178.3570 (FDA 2013) list. Excessive use of lubricants or lubricants manufactured with the use of unauthorized substances poses a danger to public health.

### 1.3.3 Physical hazards
Any foreign material not normally found in food and which is potentially dangerous to consumers is considered a physical hazard. There are many physical dangers that can be present in the raw materials such as hooks, fishing nets, plastic pieces, or pieces of glass and wood, or may originate from processing such as pieces of equipment and personal adornments.

The consumption of food adulterated by the presence of foreign bodies can result in serious injuries such as cuts in the mouth, throat, stomach, and intestines and broken teeth. In some situations, the outcome can be suffocation and death (FDA 2011).

## 1.4  Risks and benefits of seafood consumption

Although seafood is widely considered by the general population as a nutritional and healthy product – especially since brands started to emphasize the presence of ω-3 and its benefits – knowledge of most of the risks associated with seafood is not so common. Another interesting aspect of seafood, and even less known, is its contribution to the evolution of the human brain.

### 1.4.1 Seafood at the beginning of modern human brain
Before exploring the balance between the most common/measurable benefits and the risks associated with seafood, it is important to discuss what could be referred to as 'The Original Benefit'. It is known that *Australopithecus* spp. evolved

over 3 million years, but despite that it had a brain volume of about 500 cm$^3$ and almost no increase in the encephalization quotient (EQ), that is, the quotient between brain and body size. Conversely, *Homo sapiens* had an estimated brain volume of 1250 cm$^3$ and in just 1 million years, between *Homo erectus* and *Homo sapiens*, the EQ doubled (Crawford *et al.* 1999). How can this sudden increase be explained? In fact, the evolution of the brain was not vital for survival, and actually made it dependent on energy and some minerals. Even today there is still a large percentage of the population that do not consume sufficient vital nutrients for brain development. What factor could be responsible for a nonlinear Darwinistic evolution where natural selection acts to preserve and accumulate small advantages?

The fact that the earliest *Homo sapiens* individuals found in Africa are associated with lake and marine environments, and the fact that they exploited those shorelines as an accessible food source, could be involved in the genesis of the encephalization sudden growth. It is now known that the human brain is dependent on docosahexaenoic acid and requires high energy consumption, especially in infants when, at the term of birth, 74% of the energy intake is used by the growing brain (Cunnane & Crawford 2003). A diet where seafood was included could not only provide the energy and nutrients necessary for brain growth but also reduce the obligation of hunting or making of complex tools, benefiting those less fit to perform these tasks. This way, the early hominids could divert more energy from travelling in search of food to brain and other activities such as social interaction, making tools, or developing strategies to kill prey (also made possible by the increasing sophistication of the brain).

### 1.4.2 Benefits and risks

The main constituents of fish, the health benefits that consumers can enjoy by its consumption, and the risks that they are exposed to were discussed in Sections 1.1 and 1.3. The exposure to these risks/benefits is widely dependent not only on the amount consumed, but also on the particular species and size of the fish and even on the production method (harvest/aquaculture). In recent years, several research articles have been published based on risk or benefit assessment of seafood consumption in order to support or contradict the recommendations of national food safety agencies, generally encouraging seafood consumption (Sidhu 2003; Cohen *et al.* 2005; Mozaffarian & Rimm 2006; Torpy 2006; Dovydaitis 2008; FAO & WHO 2011; James 2013).

In most studies the balance was between the benefits of docosahexaenoic acid (DHA) and eicosapentaenoic acid (EPA), and the risks of methyl mercury and/or dioxins. The challenge in performing quantitative analyses is to define how much each benefit/risk affects the human health; when both negative and positive effects are balanced, an overall positive or negative result can therefore be defined. For instance, Cohen *et al.* (2005) expressed the effects in terms of quality-adjusted life year (QALY), a measure of health damage that takes into account changes in longevity and quality of life. A set of five scenarios was considered, and in only

**Table 1.3** Number of fish servings per week recommended by different food safety agencies

| Food Safety Agency | Country | Servings/week | Year recommendation made |
|---|---|---|---|
| SACN/COT | UK | General population: 4. Women in reproductive age, pregnant women and children under 16: 2 (1 serving/1 oily fish) | 2004 |
| EFSA | EU | General population: 1–2 up to 3–4 during pregnancy. Pregnant women: avoid some species with high mercury levels | 2014 |
| FDA/US EPA; AHA | USA | 2–3 | 2014 |
| Health Canada | Canada | 2 | 2007 |
| FSANZ; AHF | Australia; New Zealand | 2–3 of most types of fish. Exception for some types of fish with high levels of mercury, especially for pregnant women. | 2011 |
| MHLW | Japan | Limit consumption of some species, especially for pregnant women and children | 2005 |
| WHO/FAO | – | 1–2 | 2008 |

one there was a recommendation to reduce seafood consumption during the years of child-bearing age.

More recently, the FAO Globefish Research Programme (James 2013) released the Risks and Benefits of Seafood Consumption report, which concludes that proven benefits of seafood consumption far outweigh the possible risks. Once again, the benefit of DHA+EPA *versus* the risk of methyl mercury and dioxins was compared. A methodology was developed using numerical indices that quantified the intelligence quotient points gained/lost by children from the effect of DHA+EPA/methyl mercury, respectively. In the case of dioxins, the effect in the increase of cancer risk associated with contaminated seafood intake was quantified and compared with the benefits of DHA+EPA in mortality reduction from coronary heart disease. In both cases, seafood was classified according to its content of EPA+DHA, methyl mercury, and dioxins, and for each possible combination the positive and negative effects were compared. Only in cases of excessive consumption of a few species did the risks exceed the benefits.

Overall, none of the food safety agencies discourage fish consumption. For the biggest international organizations in the world, the total number of fish portions recommended varies from one to four portions per week with two portions per week being the most consensual number. The number of fish portions/week recommended by the principal food safety agencies are listed in Table 1.3.

# CHAPTER 2

# Food safety

## 2.1 Introduction

*Food safety* is defined by the *Codex Alimentarius* Commission as the assurance that food will not cause harm to the consumer when it is prepared and/or eaten according to its intended use. This concept should not be confused with *food security*, related to all people having access to sufficient food to meet their dietary needs. Another common expression is *food quality* or simply *quality* which is another completely different concept. It may be a little confusing at first, but quality is a relative concept. Quality is the perception that the consumer has of a series of attributes of the product, according to her/his own scale of appreciation. These attributes can be manifold. It is very common to relate food quality to the sensory characteristics of the product, since these are the attributes that are more instinctively perceived by the senses at the time of purchase or consumption. But even for characteristics such as color, odor, and taste, the appreciations can be different from one consumer to another. These appreciations can vary for reasons such as personal taste, education/area in the world where the consumer lives, or even the level of information available to the consumer. A paradigmatic example of the latter is when there is a specific color of a product with which consumers associate better quality. However, if consumers find out that that color is obtained artificially, they will probably change their mind and choose a product with another color (see Box 2.1).

Although it is common to think that concerns with guarantees of food safety are a recent subject, the health and wellbeing of consumers was being taken into consideration in ancient civilizations such as the Assyrians and the Egyptians. Records of that era report that, in order to protect the health of consumers, government authorities enforced certain rules to prevent fraudulent practices in

*Food Safety in the Seafood Industry: A Practical Guide for ISO 22000 and FSSC 22000 Implementation*,
First Edition. Nuno F. Soares, Cristina M. A. Martins and António A. Vicente.
© 2016 John Wiley & Sons, Ltd. Published 2016 by John Wiley & Sons, Ltd.

**Box 2.1** The color of salmon

The color *salmon* was named precisely after the color of salmon flesh; people therefore expect this fish to be salmon-colored. However, the actual color of salmon may in fact vary from almost white to deep pink, depending on their diet. It is very common for wild salmon to have that unique pinkish-orange shade due to the fact that its diet consists of crustaceans filled with carotenoids. On the other hand, farmed salmon, whose flesh is typically grayish-white, may obtain a similar color from synthetic carotenoids added to its feed.

food trade. Later in Ancient Greece, the purity and state of conservation of some alcoholic beverages such as wine and beer started to be inspected on a regular basis (FAO & WHO 2006b).

With the industrial revolution there was an increase in the production of large-scale consumer goods in bigger manufacturing units, which led to a faster delivery of products to people who lived far from the production centers. The 'transport revolution' contributed to this goal, associated with technical innovations in the context of preservation of perishable goods such as the introduction of mechanical refrigeration in vehicles, allowing the distribution of food to and from virtually anywhere in the world.

The concentration of food production operations in larger installations, with greater productive capacity, and distribution over large geographical areas, increased not only the likelihood of foodborne disease outbreaks (Table 2.1) but also the severity of the consequences, given the large number of consumers that began to be exposed. If we add to this the fact that these phenomena began to be massively reported by the media, especially since the 1950s and 1960s of the last century, the growing sense of insecurity among consumers towards the food that they consumed was not unexpected; such insecurities have pressed the government agencies to take measures to ensure food safety.

These measures are particularly relevant to ensure the commitment of manufacturers in the production of safe food products; in most cases, it is almost impossible to identify an unsafe product before purchase or consumption. Further, because the effects are sometimes cumulative, it is only in the long term that the consumer will suffer the consequences of ingesting unsafe products. It is essential that all those involved in the food industry ensure the safety of products regardless of the costs incurred (Mensah & Julien 2011). Due to these financial implications, it is critical to perform an evaluation and establish/reinforce evidence-based national food safety control programs that are able to fulfill the objective of having safe food products at the time of consumption, without imposing undue costs to organizations.

Regarding international trade, agreements such as the Agreement on the Application of Sanitary and Phytosanitary Measures (SPS Agreement) within the World Trade Organization are fundamental to ensure greater harmonization in standard requirements between the authorities from different countries. As

**Table 2.1** Examples of food safety issues that occurred in the last 20 years

| Year | Country | Occurrence |
| --- | --- | --- |
| 1994 | Belgium | Hormones in bovine meat |
| 1996 | Scotland | *E. coli* O157 in hamburgers |
| 1996 | UK | Antibiotic residues in meat from swine |
| 1996 | France | Clenbuterol in bovine livers |
| 1999 | France | Coal residues in Coca Cola |
| 1999 | Belgium | Dioxins in pork and dairy products |
| 2000 | Austria | Antibiotics in shrimp |
| 2006 | Portugal | Avian flu |
| 2006 | UK | Benzene in carbonated beverages |
| 2008 | China | Melamine in dairy products |
| 2009 | USA | *Salmonella* spp. in pistachios |
| 2010 | UK | Bovine spongiform encephalopathy in bovine beef |
| 2011 | Taiwan | Phthalates in food and beverages |
| 2012 | USA | *Salmonella* spp. in peanut butter |

indicated in the *Final Act of the Uruguay Round of Multilateral Trade Negotiations* signed in Marrakesh (Morocco) on 15 April 1994, each state has the right to set its own standard as long as it is based on scientific data, applied only to the necessary extent to protect the life and health of humans, animals, and plants, and as long as it does not discriminate arbitrarily or unjustifiably states in the same or similar conditions. By fulfilling these premises, the possible use of food safety as a barrier to free trade is limited; however, it also attempts to prevent distortions in competition, since unsafe products should not compete with safe products.

The consequences of ingestion of unsafe products are severe and broad. Most people have the misconception that only the less-developed countries are affected. It is in fact estimated that in developing countries 2.2 million people die each year from food and waterborne diarrheal diseases (specially children), but in developed countries up to one-third of the population may be affected each year by foodborne diseases (FAO & WHO 2006a).

It is important to distinguish that, regardless of chemical, physical, or biological origin of the hazards, these can have natural or artificial (introduced by man or by its activity) sources. When hazards originate from natural sources, it means that contamination has occurred before human intervention; in other words, a health risk is present as a result of the natural constitution of the product or exposure to a contaminant already present in its habitat. Examples of this situation are:

1. the presence of heavy metals originating from nature, regardless of those originating by human intervention;
2. the presence of microorganisms which perform, in some cases, vital functions during the life of the animal;
3. the presence of support structures, such as bones or shells.

On the other hand, hazards of artificial origin result both from the processing and handling of the product by humans before its consumption, and from the introduction of substances into the product as a result of human activity which eventually contaminate food.

This difference is very relevant when deciding how to control food hazards. Generally, the approach to natural hazards is to verify whether they are present within an acceptable limit or to act towards their containment; in the case of artificial hazards, most times there is the possibility to act at their source in an attempt to ensure that they are absent from the food.

## 2.2 The Codex Alimentarius

FAO/WHO *Codex Alimentarius* consists of a set of standards, internationally recognized and based on scientific knowledge, which arose from the need to harmonize food standards and which began to proliferate in several countries at the beginning of the last century. Its name, which is Latin for 'food code' or 'food law', and some of its organizational aspects derive from the *Codex Alimentarius Austriacus*. This code consists of a set of standards and product descriptions which were developed between 1897 and 1911 for food in the Austrian–Hungarian Empire that, despite having no legislative authority, was considered a reference in courtrooms (Davies 1970).

The key steps for the elaboration of the FAO/WHO *Codex Alimentarius* were as follows.

- October 1960: First conference of FAO for Europe, during which the need to establish an international agreement on food standards to ensure the safety of food and facilitate international food trade was recognized (FAO 1960).
- November 1961: At the eleventh session the resolution that originated the *Codex Alimentarius* Committee (CAC) was approved (FAO 1962).
- June 1962: The cooperation between FAO and WHO was formalized during the Joint FAO/WHO Food Standards Conference.
- October 1963: First meeting of the *Codex Alimentarius* Commission.

Other significant dates include:

- 1969: the General Principles of Food Hygiene were adopted;
- July 1993: the importance of hazard analysis and critical control points (HACCP) in food safety control was recognized and a guide for its implementation was published; and
- June 1997: the guidelines for HACCP implementation were included in the annex of the General Principles of Food Hygiene's third revision.

The three factors that most contribute to the credibility of the CAC are a solid scientific basis, a global membership, and the fact that, when the adoption or amendment of standards is necessary, a demand of consensus among its members is mandatory (FAO & WHO 2015). According to the information present in the

*Codex Alimentarius* website[1], 99% of the world population is represented among the 186 members of the committee (185 member countries and 1 member organization, the European Union) and 229 observers (52 intergovernmental organizations, 161 nongovernmental organizations and 16 United Nations organizations).

Although the application of *Codex Alimentarius* is not binding from a legal point of view, its reputation makes it a reference both to the development of national legislation or Food Safety Management Systems, and to the resolution of international trade disputes as a benchmark (used by the WHO) in compliance with the SPS Agreement.

The *Codex Alimentarius* consists of a set of various documents including, among others, Standards, Guidelines, and Codes of Practice. From the *Codex Alimentarius* website[2] it is possible to conclude that the majority of the publications are Standards (63%), followed by Guidelines (21%) and Codes of Practice (14%). The first code published by CAC was the Recommended International Code of Practice – General Principles of Food Hygiene in 1969, which specified the essential principles of food hygiene applicable at all stages of the food chain from production to final consumption. The current version of this code also includes (in Annex) the HACCP and the guidelines for its implementation, adopted by CAC in 1997 (FAO 1998).

## 2.3 HACCP system: Hazard analysis and critical control points

### 2.3.1 History

HACCP was developed by The Pillsbury Company working alongside NASA and the US Army Laboratories in the 1960s in order to develop products that were safe for astronauts, since it was very important to minimize the chances of them becoming ill during space missions, potentially compromising its success. One of the foods developed was an energy bar. This product was associated with the origins of HACCP application in the food industry (Wallace *et al.* 2011; Box 2.2; Fig. 2.1).

**Box 2.2**   Space Food Sticks

Space Food Sticks were a 'high-energy snack' commercialized by Pillsbury Company and publicized in a 1970s Washington Post ad as a 'nutritious good-tasting snack developed by Pillsbury for long space flights'. These energy bars were, for a brief period of time, available to consumers in food stores (Fig. 2.1). Space Food Sticks disappeared from North American supermarket shelves in the 1980s. They were later recreated by Retrofuture Products and re-launched in October 2006 in two flavors: chocolate and peanut butter.

---

[1] http://www.codexalimentarius.org/members-observers (accessed 4 September 2015)
[2] http://www.codexalimentarius.org/standards (accessed 4 September 2015)

**Figure 2.1** Space Food Sticks were designed for astronauts and later marketed to the public. Source: Wallace *et al.* (2010). Reproduced with permission from John Wiley & Sons.

NASA already had experience of identifying critical control points in engineering management. As Dr Paul A. Lachance mentioned in an interview (by Jennifer Ross-Nazzal Houston, Texas and New Brunswick, New Jersey on 4 May 2006): 'Pillsbury probably was the first to put the acronyms together. We had a hazard thing, and we had a requirement, you know, a plan document, reliability document, and we imposed that. So the CCP really came from NASA.' At the time, Dr Lachance, who was NASA Flight and Food Coordinatore, worked with Dr Howard Bauman from Pillsbury Company in adapting these concepts to the fabrication of food products for NASA.

In 1971 the Pillsbury Company had a serious food safety problem as one of its products, intended for consumption by infants, was contaminated with glass (see Box 2.3). This problem had a major media impact and the company had to withdraw the product from the market. Following this incident, Pillsbury decided to apply the same principles they had developed with NASA to guarantee safe products, and announced this decision publicly in order to regain credibility with their customers (Ross-Nazzal 2007).

Later that year, Dr Bauman gave a presentation at the National Conference on Food Protection sponsored by the American Public Health Association (APHA & FDA 1972) during which he made a description of critical control points and good manufacturing practices (although not yet using the acronym HACCP).

**Box 2.3**    From NASA to retail

In the mid-1960s Pillsbury worked with the US Army and NASA to provide safe foods for military personnel and astronauts. Their efforts gradually led to the modern HACCP system. The thought of a catastrophic 'two-bucket' illness during a space mission was a principal driving force behind their effort. In 1971 Pillsbury was jolted into action to apply the embryonic HACCP system to its retail products when a prominent national radio reporter exclaimed, 'Good morning, America. There's glass in your baby food!' Pieces of glass from a light fixture had fallen into a farina storage bin. The contaminated farina was packaged for retail sale as baby food. The Pillsbury CEO declared that such an incident would never occur again.

Source: Dr William H. Sperber

A revolution in the food industry and a paradigm shift had been initiated; the future was of a proactive and not a reactive approach to food safety problems.

In Box 2.4 Dr William Sperber gives its personal insight on the HACCP early times and future perspectives. However, many years passed until the next major step for the international dissemination of HACCP occurred. It was only in the 1990s that the National Advisory Committee on Microbiological Criteria for Foods (NACMCF) and *Codex Alimentarius* published reports on HACCP. In 1997 the CAC published a guide to its implementation, making it an essential tool for all organizations that wished to do so.

## 2.3.2 HACCP system

The HACCP system is a scientific methodology that identifies, evaluates, and controls biological, chemical, and physical hazards associated with food or that may be introduced during its processing. For this reason it is considered a preventative system which helps to reduce the need for inspections and testing of finished products (FAO 1998).

In addition to enhancing food safety, this system also presents other advantages (FAO 1998):
- increases customer and consumer confidence;
- reduces costs through reduction of product losses and rework;
- promotes international trade; and
- can be integrated with other management systems.

HACCP consists of 12 steps, of which the first 5 (shown in Fig. 2.2) have the main goals of establishing a HACCP team and collecting all the information necessary for hazard analysis. The remaining 7 steps (Fig. 2.3) correspond to the 7 principles that comprise this methodology (FAO 1998).

For the development and effective implementation of HACCP, organizations must first ensure compliance with a set of requirements, called prerequisites, which promote the conditions necessary to obtain safe and adequate products for their intended purpose. In some parts of the food chain where it is impossible to

**Box 2.4**    Interview with Dr William H. Sperber

William H. Sperber studied biological and chemical sciences at the University of Wisconsin in Madison, culminating in a Ph.D. in microbiology and biochemistry in 1969. He worked for 43 years in research and management positions with major global food companies, the majority with the Pillsbury Company which developed the HACCP system of food safety management. Dr Sperber has researched, developed, and taught the HACCP system and its prerequisite programs since 1972, quite likely the longest tenure in this aspect of food safety management. Following his retirement in 2012 he continues to advance food safety efforts as president of The Friendly Microbiologist LLC by advising individual food professionals and their companies. He remains an ardent proponent of Good Consumer Practices and the development of a global food protection organization.

**When asked by Pillsbury in 1972 to perform a hazard analysis and work with Dr Bauman, what did you find there that could be compared to what we call today of HACCP?**
The first rudimentary HACCP system had been established just before I arrived at Pillsbury. That system consisted of three principles involving hazard analysis, critical control points, and monitoring procedures.

**Was there a clear idea of what needed to be done?**
Yes, Pillsbury had created an extensive internal HACCP training program. The same training program was used in September 1972, in a three-week program to train US Food & Drug Administration inspectors, who in 1974 published regulations to assure protection against botulism in both low-acid and acidified canned food. I believe these were the first regulations of this type anywhere in the world.

**Was it clear to you and the company that you were initiating the application of a methodology that would be so significant in the future of the food industry?**
We knew our original HACCP program was necessary to improve the safety of Pillsbury's consumer foods. We did not realize then that the HACCP system would be used quickly by other US food processors, that it would mature to 7 principles and then gain global acceptance, all within 20 years.

**Which major difficulties did you find in those early stages of the project?**
As each HACCP is unique to each product type in each production plant, we learned that it was important to have adequate training of the HACCP teams at each plant, and to have continued training over the years as new employees were hired. We also learned that our HACCP system did not assure food safety by itself. It needed the support of prerequisite programs such as good manufacturing practices (GMPs). In short, the HACCP system provided assurance of the safety of the food product. GMPs provided assurance of the sanitary condition of the surrounding production environment.

**Do you still remember which hazards were identified as critical control points in the first analysis? After 40 years it seems complicated to perform a hazard analysis without access to extended and detailed information on microorganisms (and even on chemical hazards), such as the one available nowadays. Does this mean that in those days you would be mostly concerned about physical hazards?**

Once the world's largest flour miller, many of Pillsbury's products were flour based. Of necessity, there was a heavy emphasis on physical hazards such as foreign materials, broken screens, glass, etc. However, the primary emphasis on products such as canned foods had a major focus on microbiological hazards. The principal microbial hazards were known 50 years earlier.

**How do you see the future of HACCP? Do you foresee the obligation of applying it all over the world? How do you evaluate the evolution in time regarding the motivations for HACCP implementation?**

The 1992 recommendation of HACCP by *Codex Alimentarius* provided a unique opportunity to have a single global standard for food safety management. To a large extent that system is applied today, largely through the necessary efforts of global food ingredient, food product, and food service corporations. This arrangement would be even more effective if each national government would require full compliance with the Codex recommendations. I don't know how many nations have such a requirement. The US does not. Such requirements are necessary in the new global economy in which food ingredients produced in one country will be distributed to almost every country in the world.

**Do you think that most companies still implement HACCP because they need to answer to some kind of legal demand or because their clients demand them to (as it was the case of Pillsbury – they needed to regain the trust of their clients after the incident with the glass) and that, in this way, HACCP is seen more as an economic burden to them than an honest search for the safety of their clients?**

We have never viewed HACCP and food safety management as a costly burden. It is more costly to endure and correct food safety failures that undermine public confidence. The development and use of HACCP plans and prerequisite programs is the honest search for the safety of food consumers.

1. Constitute the HACCP team    2. Describe the product    3. Identify the intended use    4. Building a flow chart    5. Check the flow chart on site

**Figure 2.2** Five preliminary steps for the application of the seven principles of HACCP.

implement a HACCP plan, the prerequisites (presented in detail in Chapter 4) are the only way to control the safety of products.

As advocated by Sperber (2005) the possibility of implementing HACCP 'from farm to fork' is an illusion because it is often impossible to implement control measures or establish critical control points in stages such as primary production and final consumption. A guide developed in 2005 by the Health & Consumer Protection of the European Commission shares the same convictions (EC 2005). This document seeks to explain the reference in Recital 15 of the Reg. 852/2004 (in which it is stated that food business operators in the European

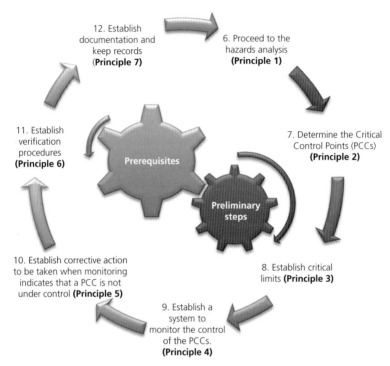

**Figure 2.3** Relationship between Prerequisite Programmes and the HACCP system steps.

Union must implement procedures based on the HACCP principles) where it is declared that:

> The HACCP requirements should … provide sufficient flexibility to be applicable in all situations, including in small businesses. In particular, it is necessary to recognize that, in certain food businesses, it is not possible to identify critical control points and that, in some cases, good hygienic practices can replace the monitoring of critical control points.

Throughout the guide a few considerations of this flexibility are discussed, of which the following stand out.

- The possibility of a simplified approach could be used as long as it does not compromise the objective of the seven HACCP principles of guaranteeing the identification and control of significant hazards.
- In some food businesses where there is no preparation, manufacturing, or processing of food (such as in marquees, market stalls, mobile sales vehicles, small retail shops, or transport and storage of pre-packed food or nonperishable food) it is possible that all hazards can be controlled through the implementation of the prerequisite requirements.
- In some sectors, generic HACCP guides could be very important to assist the implementation of procedures or methods; one of the sectors considered are establishments handling fishery products.

- The critical limit at a critical control point does not always have to be expressed as a numerical value; this is the case for procedures based on visual observation, for example.
- Record keeping is also addressed, mainly to reinforce what is clearly postulated in the Recital 15 of Reg. 852/2004, that is: '… the requirement of retaining documents needs to be flexible in order to avoid undue burdens for very small businesses.' A similar approach is given to training of personnel, which must be adapted to the size and nature of the business.

## 2.4  Food safety standards

A brief history of three standards that, as for the FSSC 22000 (see Chapter 5), were benchmarked against the GFSI Guidance Document Sixth Edition for the scope Processing of Animal Perishable Products, is described the following sections (GFSI 2015). All of these standards apply to the global market of food products, having in common the implementation of a food safety system which is based on the HACCP methodology and the *Codex Alimentarius* principles and complies with the current legislation. Box 2.5 introduces the history and mission of GFSI.

Table 2.2 provides a comparison between the number of certification bodies, certified organizations, and the predominant markets of International Features Standard (IFS), British Retail Consortium (BRC), Safe Quality Food (SQF) and FSSC 22000.

**Box 2.5**   Global Food Safety Initiative (GFSI)

---

**'Once certified, accepted everywhere'**

This was the 'The Proposal' that set the tone for the creation of GFSI back in the year 2000.

Soon after some of the most publicized food safety crises in the 1990s such as bovine spongiform encephalopathy (BSE), dioxins, and listeria outbreaks, some of the most pre-eminent world's food retailers decide to found the Global Food Safety Initiative (GFSI). In fact, it was the confluence of a very low consumer confidence in food safety (after those incidents) and the numerous standards that were emerging from retailers and brand manufactures (that imply several audits a year against each one of those standards) that provided the compelling motivation to take action.

GFSI is a nonprofit foundation with the mission to provide continuous improvement in food safety management systems to ensure confidence in the delivery of safe food to consumers worldwide. The approach taken is a benchmarking model where food safety management schemes are recognized according to defined requirements in a guidance document. This model enables not only to guarantee equivalence and convergence between food safety management systems to be guaranteed, but also that organizations still retain control to choose between different schemes (GFSI 2013).

Despite 15 years passing since the 'The Proposal', it has not yet been fully accomplished. A study intended to assess the efficacy and business impact of the implementation of GFSI-recognized schemes, **GFSI Efficacy Research Project** (GFSI & Sealed Air, 2014),

demonstrates other major benefits:
- 61% considered it increases the ability to produce safe food;
- 72% considered it enhances company food safety practices; and
- 90% considered it raises employee awareness and knowledge.

Other interesting conclusions include the following.

1. The major drive to implement a GFSI-recognized scheme is external: To the question of what is the most important factor leading to certification, 67% responded 'existing customer requirement, to continue doing business with them'. The second-most common (9%) answer was 'potential new customer requirement, to start doing business with them'.
2. The respondents tended to disagree that certification led to less work (31%), compared to 24% who agreed. These tendency is even stronger when the question was reducing the working capital loss, where 'disagree' surpassed 'agree' by 28%.
3. There were a significant number of respondents (67%) who identified an increase in the hours spent on internal auditing and on production worker training. Nevertheless, 87% said that it had been very or fairly beneficial to be certified against a recognized scheme.
4. Regarding the implementation length, 64% indicated that the entire process took less than a year.
5. Food safety issues were considered to be more regularly scheduled on the agenda of senior managers by 61% of the respondents. These percentages increased even further (68%) when the involvement of top leaders in food safety was queried.
6. An increase in annual sales/revenue was reported by 45% of the respondents (7% declared decreasing results).

**Table 2.2** Comparison between the number of certification bodies and certified organizations and the predominant markets from IFS, BRC, SQF, and FSSC 22000

| Schemes | Year of first publication | Predominant markets | Certified organizations | Certification bodies[a] |
|---|---|---|---|---|
| SQF | 1994 | North American and Australian | 6,745[a] | 32 |
| BRC | 1998 | British | 22,550[a] | 111 |
| IFS | 2003 | German, French, and Italian | 17,235[b] | 103 |
| FSSC 22000 | 2009 | European and Asian | 11,338[a] | 107 |

[a] Numbers obtained from schemes website on 1 June 2014
[b] Number of audits performed during 2014 under the IFS umbrella

## 2.4.1 IFS Food Standard (Version 6)

The third version of the IFS Food Standard (the first version which was implemented) was developed by the associated members of the German retail federation – *Handelsverband Deutschland* (HDE) – and published in 2003. The update of this document (version 4) in January 2004 was the result of a joint effort between the HDE and the *Fédération des Entreprises du Commerce et de la Distribution* (FCD). Since then, and up until version 6 (last updated in April 2014), the development of this standard was extended to the participation of Italian retail associations (which joined the IFS), working groups, several retailers from Switzerland,

Austria, Spain, Asia, North and South America, and to other stakeholders (industry, food services, and certification bodies) (IFS 2014).

The IFS Food Standard is especially important for manufacturers of retailer branded food products, and aims to assess the food safety system and the quality of the suppliers. It is only applicable to processed foods or in cases which there is the risk of contamination during primary packaging. The IFS also developed other standards specific to other areas of the food chain such as brokers, cash and carries/wholesales, food stores, and logistics (IFS 2015).

### 2.4.2 BRC Global Standard for Food Safety (Issue 6)

In October 1998 the British Retail Consortium (BRC), alongside many distributors, launched the Technical Standard and Protocol for Companies Supplying Retailer Branded Food Products. As the name implies, the standard was originally developed for manufacturers supplying retailer brand products (Arfini & Mancini 2014).

The evolution of the standard and its increasing use outside the UK led the BRC to change its name in order to get the scope across more effectively. It was therefore renamed to BRC Global Standard Food in January 2003. As stated in issue 7 of the standard, '… the standard provides a framework for food manufacturers to assist them in the production of safe food and to manage product quality to meet customers' requirements.' Currently, the standard defines requirements 'for the manufacture of processed foods and the preparation of primary products supplied as retailer-branded products, branded food products and food or ingredients used by food service companies, catering companies and food manufacturers' (BRC 2015). BRC has also developed other standards in areas such as packaging, consumer products, storage, and distribution.

### 2.4.3 SQF Code (7th Edition Level 2)

SQF stands for Safe Quality Food and, since 2013, is owned by the Food Marketing Institute (FMI) and managed by the Safe Quality Food Institute.

This code was first developed in Australia and in 1994 released the SQF 2000 for food industry and later in 1998 the SQF 1000 for primary production. In 2012 the SQF 2000 edition 6 and SQF 1000 edition 5 were remodeled into a new SQF code, intended for use by all sectors of the food industry (e.g., primary production, transport, retailing, manufacture) (SQF 2008, 2010, 2014b).

# CHAPTER 3

# The EN ISO 22000:2005

## 3.1 History

The purpose of an organization should be to ensure that their products can be consumed without constituting a health risk. The variety of safety levels and the requirements of national legislation have made it increasingly difficult to achieve this objective in a world where food products are traded around the globe and the food chain is increasingly complex. The failure of one organization in the supply chain can make the product unsafe and therefore constitute a risk to consumer health.

As a result of the work of the Technical Committee ISO/TC 34, the ISO Standard 22000:2005 *Food safety management systems: Requirements for any organization in the food chain* was officially published in September 2005. This committee met not only with delegates of the member countries but also with other prominent organizations, such as the *Codex Alimentarius* Commission, Confederation of Food and Drink Industries of the European Union (CIAA, later renamed to FoodDrinkEurope), and WHO. The aim was to prepare a reference, based on the draft standard proposed in 2001 by the Danish Association of Standardization, that could structure the principles of the HACCP transversely across the industry and be applied worldwide (ISO 2008a).

The most recent numbers from companies certified by this management system show a continuous growth since 2007, reaching more than 25,000 organizations in 2013 as shown in Figure 3.1.[1] Despite the growth, the statistical analysis shows that a substantial number (83.0% in 2013) of the organizations is located in Europe and East Asia and the Pacific.

---

[1] These data are based on voluntary registrations made by companies and are therefore not exhaustive.

---

*Food Safety in the Seafood Industry: A Practical Guide for ISO 22000 and FSSC 22000 Implementation*, First Edition. Nuno F. Soares, Cristina M. A. Martins and António A. Vicente.
© 2016 John Wiley & Sons, Ltd. Published 2016 by John Wiley & Sons, Ltd.

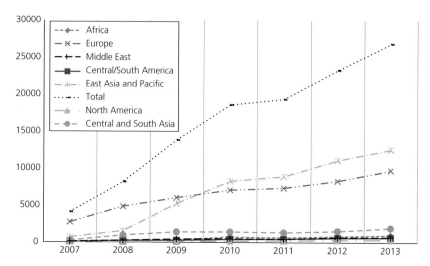

**Figure 3.1** Number of certificates reported each year by individual certification bodies worldwide (based on data available in ISO website, http://www.iso.org/iso/iso-survey).

## 3.2 Structure

The ISO 22000:2005 was structured in order to be applied to any organization operating in the food industry, by promoting the 'accountability' of each operator for the guarantee of the product safety. This characteristic constituted an added value but was also responsible for what is considered its main limitation, the prerequisites being too generic.

The standard specifies the requirements necessary to implement a food safety management system that ensures the safety of food throughout the food chain (ISO 2005a). However, in the standard there are no concrete guidelines on how to implement them. Each organization will need to assess how best to ensure compliance according to their specific situation and position in the food chain. It is clearly stated that the food safety management system must guarantee compliance with all statutory and regulatory requirements and promote interactive communication throughout the food chain and the continuous improvement of the system.

The standard is structurally very much in line with ISO 9001, facilitating its integration in organizations that have ISO 9001 already implemented. In fact, both standards can be divided into three main sections: introduction; management requirements; and (safe) product realization, validation, verification, and continuous improvement. Table 3.1 presents the clauses (in the case of ISO 22000:2005) for each of the three main sections.

Clauses 1–3 are considered not auditable, as they show no guidelines for the implementation of the standard. In clauses 4–6 the integration of food safety management in a general framework of the organization's management is promoted.

**Table 3.1** Clauses of ISO 22000:2005 included in each of the three main sections proposed

| Section | ISO 22000:2005 |
| --- | --- |
| Introduction | 1. Scope |
| | 2. Normative references |
| | 3. Terms and definitions |
| Management Requirements | 4. Food safety management system |
| | 5. Management responsibility |
| | 6. Resource management |
| (Safe) Product Realization, Validation, | 7. Planning and realization of safe products |
| Verification, and Continuous Improvement | 8. Validation, verification, and improvement |
| | of the food safety management system |

The 7 principles and 12 steps of the implementation of the HACCP system are encompassed in Clause 7 of the standard with the exception of the requirements regarding documentation, which are also included in Clause 4. Clause 8 focuses on the evaluation of the ability of the system to guarantee safety as well as its continuous improvement.

## 3.3 Implementation

### 3.3.1 Pressure (drivers)

There are many factors that pressure organizations to implement a food safety management system. They can be divided into either internal or external pressures, described in the following paragraphs.

Internal pressures (usually those with the least strength), include the decision of the management to either optimize food safety procedures or to reduce costs with noncompliant or returned products. In the latter case however, it's unclear whether (in general) the reduction in costs is greater than the investment necessary for the implementation, especially for small organizations.

External pressure refers to the implementation of a food safety system such as ISO 22000, BRC, IFS, or SQL which, although described as voluntary for the organizations, may be mandatory in order to keep existing customers and obtain new ones. When customers (particularly large retailers) only accept suppliers certified by a Food Safety Management System (FSMS) or when its application is widespread in the competition, there is a great commercial pressure to the effect that its implementation is no longer voluntary. This requirement can often be seen as a barrier to free trade in food products, even among countries with trade agreements, preventing the access of enterprises from less-developed countries to food distribution chains. Consumer protection groups and the media can also be a pressure factor, as comparative testing results of food products can damage the impression that consumers have of the products. Finally, there is pressure from the individual consumer thanks to the use of social networks. Since the HACCP and ISO 22000 systems aim to prevent unsafe products from reaching the consumer, their implementation can avoid consumers' bad experiences of a certain product. Such unpleasant experiences would previously have come to the attention of just those people closest to the consumer, but today such experiences can have much wider consequences due to the rapid and far-reaching impact of social networks.

Another important pressure is the avoidance of negative publicity from fines or even criminal prosecution as a result of a foodborne poisoning outbreak. Box 3.1 presents an interview on these subjects with Bill Marler, a foodborne-illness attorney.

**Box 3.1**    Interview with Bill Marler

Bill Marler is a graduate from the Seattle University School of Law class of 1987. In 1998 he became the Law School's 'Lawyer in Residence.' Mr Marler received undergraduate degrees in Political Science, English, and Economics from Washington State University in 1982. While attending WSU, he was elected to the Pullman City Council. At the age of 19, he became the youngest person (and the first student) ever elected. Mr Marler received the Distinguished Achievement Award from the WSU College of Liberal Arts in 1997. In 1998, Governor Gary Locke appointed Mr Marler to the University Board of Regents. He recently served as President of the Board. He also served on the State Higher Education Coordinating Board. In April 2010 Mr Marler was awarded the prestigious NSF Food Safety Leadership Award for Innovation in Education. In 2008, he was awarded both the Seattle/King County Bar Association 2008 Outstanding Lawyer Award and the Washington State Trial Lawyers Association's 'Public Justice Award,' and in 2013 he was honored with the Seattle University Distinguished Law Graduate Award. The attorneys of the State of Washington have chosen Mr Marler as a 'Super Lawyer' every year since 1998. He has an AV rating from Martindale–Hubbell, and is listed in the Bar Register of Pre-eminent Attorneys. He has been chosen as one of 'America's Best Lawyers' every year since 2009. He is married to Julie Marler and they have three daughters, Morgan, Olivia, and Sydney.

**In your opinion, are the recent criminal convictions in the USA punctual or part of a global trend to hold top management responsible in cases of food poisoning? Do you have information about similar cases in other countries, where the managers have been convicted and sent to prison?**
In the 22 years that I have been working on food cases since the Jack-in-the Box *E. coli* outbreak back in 1993, there have been a handful of criminal convictions of companies involved in food outbreaks. Let me give you some examples.

In 1998, in what was the first criminal conviction in a large-scale food-poisoning outbreak, Odwalla Inc. pleaded guilty to violating Federal Food Safety laws and agreed to pay a $1.5 million fine for selling tainted apple juice that killed a 16-month-old girl and sickened 70 other people in several states in 1996. Odwalla, based in Half Moon Bay, California, pleaded guilty to 16 counts of unknowingly delivering 'adulterated food products for introduction into interstate commerce' in the October 1996 outbreak, in which a batch of its juice infected with the toxic bacteria *E. coli* O157:H7 sickened people in Colorado, California, Washington, and Canada. Fourteen children developed a life-threatening disease (hemolytic uremic syndrome, HUS) that ravages the kidneys. At the time, the $1.5 million penalty was the largest criminal penalty in a food poisoning case. Odwalla was also on court-supervised probation for five years, meaning that it had to submit a detailed plan to the Food & Drug Agency demonstrating its food safety precautions; any subsequent violations could have resulted in more serious charges.

In 2013, Austin 'Jack' DeCoster and his son, Peter DeCoster, both faced charges stemming from a *Salmonella* outbreak caused by their Iowa egg farms in 2010. The *Salmonella* outbreak ran from 1 May to 30 November 2010, and prompted the recall of more than a half-billion eggs. While there were 1939 confirmed infections, statistical models used to account for *Salmonella* illnesses in the US suggested that the eggs might have sickened more than 62,000 people. The family business, known as Quality Egg LLC, pleaded guilty in 2015 to a federal felony count of bribing a USDA egg inspector and to two misdemeanors of unknowingly introducing adulterated food into interstate commerce. As part of the plea agreement, Quality Egg paid a $6.8-million fine and the DeCosters $100,000 each, for a total of $7 million. Both DeCosters were sentenced to three months in jail. They are appealing the jail sentence.

In 2014 former Peanut Corporation of America (PCA) owner Stewart Parnell, his brother and one-time peanut broker, Michael Parnell, and Mary Wilkerson, former quality control manager at the company's Blakely, Georgia plant faced a federal jury in Albany, Georgia. The 12-member jury found Stewart Parnell guilty on 67 federal felony counts, Michael Parnell was found guilty on 30 counts, and Wilkerson was found guilty of one of the two counts of obstruction of justice charged against her. Two other PCA employees earlier pleaded guilty. The felony charges of introducing adulterated food into interstate commerce, 'with the intent to defraud or mislead,' stemmed from a 2008–2009 *Salmonella* outbreak that sickened 714 and left 9 dead. Stewart has recently (September 2015) been sentenced to 28 years in prison, his brother Michael 20 years and quality control manager Mary Wilkerson 5 years.

I think that these criminal prosecutions are an attempt by the FDA and the US Department of Justice to make an example of companies that produce unsafe food. You might have more success in prosecuting organizations that do not guarantee food safety if the Food Safety Modernization Act is fully founded and the number of government inspectors are adequate, but this has not yet been the case. I don't conduct a lot of research regarding other countries, but there have been similar convictions elsewhere.

**You recovered over $600 million in compensation for your clients. Have you noticed if, in the last years, top management has been motivated to comply with food safety, knowing the financial consequences and even the risk of facing jail time that can result from these lawsuits?**

I think so. I speak in conferences around the world and one of the main topics that people want me to talk about is the criminal aspect and if [top management] can be at risk (fines or even jail). I do think that because there are more convictions and prosecutions in the last five years, [food safety] is certainly something that people are paying attention to.

**How do you see the role and responsibility of third-party audit organizations in guaranteeing food safety? Will the recent tendency for unannounced audits make a difference?**

To fully understand the problem with third-party audits we must first have a historical perspective. In 1906, under the first Roosevelt Administration, the government had inspectors in every meat plant that, despite the fact they were supposed to be paid by industry, were government employees. With the growth in the food industry, especially after World War II, the FDA become unable to oversee so many organizations from the other sectors of food industry, producing and distributing many goods such as fish, vegetables, fruits, cheese, and grain, as well as imported goods all over the country. The problem was that this growth was not matched by growth in an infrastructure of inspectors. It is common that the FDA only inspects organizations once every three years. For instance, in the Peanut Corporation of America case, they had not been inspected by the FDA in the 5 years before the outbreak. Part of the problem is that there is not appropriate funding for enough inspectors.

What industry have done, because they are concerned with who their suppliers are, is help create those third-party audits companies to fill the blank left from the government not providing enough inspectors to regularly monitor food producers. In my experience, the big companies require these third-party audits of their suppliers, but they rarely analyze the results of the audit (as long as they pass). Furthermore, the suppliers pay the auditors and so a clear conflict of interest is created. For that reason, it is not a surprise that in many of the food illness outbreaks, companies had been audited and achieved great results.

Auditing should be performed for the benefit of consumer, and if it is not possible for a government inspector to perform the audit then it should be the responsibility of the big companies/retailers. They should at least pay for the audits to guarantee the absolute independence of the auditor.

I don't know if we can say that inspectors should be held responsible for food safety, since it has to be guaranteed by the producers, but I think that third-party audit companies could ensure the total impartiality of the auditors. I think that incentives for third-party audits are wrong; the millions of dollars spent on them could be used to reinforce the ability of the FDA to perform the inspections.

**In your 22 years of experience you have seen the introduction of all the food safety systems SQF, BRC, IFS and, more recently, ISO 22000 and FSSC 22000. Have you ever had a case where the company was certified by one of these schemes? Do you think that organizations certified by these are more prepared not only to guarantee food safety but also to defend themselves from prosecution when a problem occurs?**
Fully implemented food safety schemes, based on HACCP with audits and commitment to food safety, and schemes that are actually followed and not just described on paper, reduce the chances of the organization having problems. If something happened despite these precautions, they still have the HACCP in place and are doing all the things they are supposed to do. In many countries that is a defense: 'due diligence.' In the US we do not have this but, if the companies can show that they were trying to do the right thing, it may reduce the consequences.

## 3.3.2 Method of implementation

The way in which an organization implements the standard is conditioned by its size, position in the food chain, activity, and the resources available. There are two main approaches to implementation:

- using internal human resources; or
- using external consultants.

An important assessment, particularly for the first approach, is to verify whether the internal human resources have the training/education required for the implementation or if it is necessary to strengthen that knowledge. These skills need to be evidenced, especially when organizations intend to obtain certification.

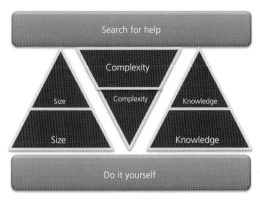

**Figure 3.2** Key factors for determining whether an organization should seek external expertise.

The larger the company, the greater is the likelihood that they will be able to implement the system without the help of external consultants (Fig. 3.2). The complexity of the activity and the level of product handling are important factors to be taken into account. For example, the implementation of a standard in a company that does not manipulate the product is simpler; it is therefore easier to dismiss external support. Finally, if the company already possesses human resources with training in the food/FSMS area and are aware of the processes, it may well dismiss external consultants. This subject is also approached in the definition of the Food Safety Team (see second part of Section 4.5.3) and Food Safety Team leader (see Section 4.3.5).

### 3.3.3 Difficulties and challenges

Many difficulties and setbacks must be overcome in order to implement and maintain ISO 22000. The most obvious, and the challenge which must be dealt with first, is the inherent costs of implementation. As previously mentioned, organizations must assess the ability of their human resources to implement the process, which may involve the need for external training, recruiting consultants, or even new employees. Organizations will then have to consider the possibility of making structural changes to facilities or acquiring new equipment. If the activity of the company is carried out in rented facilities, they may have additional difficulties in making structural changes.

Many other costs will be incurred later with the implementation, maintenance, and certification of the standard, including monitoring processes, analysis of products in accredited laboratories, external audits fees, and training.

The main difficulties to be overcome are usually not financial but related to human resources, however. Very often there is a resistance to changing behaviors and procedures; in contrast to what might be expected, these objections do not have their origin in production operators. Another common resistance is the 'lack of time.' There is commonly a greater need to find time for meetings, preparation of documents, and training, especially in the implementation phase. For this reason, it is very important that all employees are involved, especially the top management; as we shall see in Section 4.3, the top management must be committed to the whole process. In organizations where there is a high turnover of staff or a large use of temporary work, all these problems are aggravated.

### 3.3.4 Benefits

The benefits from implementing the standard to the organization are not always tangible or easily measurable. More direct benefits, such as the reduction of product failure during manufacture and the reduction of discards, are easier to identify and quantify. What is much more complex to evaluate in monetary terms is the impact of standard implementation on: the image of a company (and if this has consequences in terms of business volume); reduction of litigation cases or fines for noncompliance with statutory requirements; and the professional development and awareness of all employees regarding issues of food safety.

The implementation of a FSMS which is periodically evaluated by external auditors facilitates, in many cases, the work of those who have to establish rules for the employees. The fact that employees will be regularly evaluated by people from outside the organization will make them more willing to comply with the rules. It will also make them understand that those rules are not just specific to them, but in fact are principles common to all food businesses. Another benefit of being audited by external entities is that the recommendations and advice of the auditors can provide input for improvement of the FSMS, even in non-food-safety-related processes.

Finally, a benefit of standard implementation may be entry to new markets. When organizations are faced with the need to implement an FSMS in order to supply a particular customer or have access to a new market, the implementation of the standard is no longer voluntary in reality. For this reason this is not a benefit that is normally considered in the trade-off between the costs/benefits of implementing the standard, instead being considered as a business essential.

## 3.4 Changes in the organization

### 3.4.1 Initial resistance

The first impact of the implementation process of the standard is usually very negative. Many people, especially if they cannot engage and motivate themselves, will consider that much human and financial resources are being spent on the project. Even if the organization does not have to recruit new employees or consultants, it will always be necessary to provide time for meetings and most likely financial resources for training and equipment.

Adding to this scenario, the first impact of this phase for most employees results in:
- changes of procedures they were used to;
- the need to complete new records;

- the need to attend demonstrations, often given by people from outside the company, on common-sense actions such as hand washing.

The above situations might simply represent additional work to some. There are ways to reduce this initial resistance, but the most important is the effective commitment of the management. As shown in Section 4.3, the standard has a clause (5) fully dedicated to the responsibility of management; in Clause 5.1 it is clearly indicated how the management should demonstrate their commitment. It is very important that everyone recognizes the commitment of the management to the project which involves providing resources and, especially, leading by example. Seeing the top management visiting the factory wearing hair nets, attending the same awareness training for food safety, or participating and developing the system documents has a great impact on the motivation of all to implement the standard. One of the most common factors that employees identify as a reason for noncompliance of the principles of the standard is the lack of training, commitment, and exemplary attitudes by the top management.

Another way to improve receptivity to change is, in the early stages, implementing some changes that have a positive impact on the lives of employees such as: automating or simplifying the filling of records; altering the processes to minimize food handling; undertaking, when necessary, interventions to improve employee facilities and service areas; improving tools and equipment to be used during cleaning and sanitizing; acknowledging the importance of employees in the implementation of the system (meeting with employees to hear their opinions/suggestions); and improving knowledge of employees by providing training programs.

Finally, visiting other companies that already have the system working, not only to see how the system can produce results but also to discuss how difficulties were overcome, can motivate employees to the project.

### 3.4.2 Other changes

Other consequences are common in organizations that implement the standard. As mentioned in Section 3.3.2, it may be necessary right from the start to recruit new employees to lead the process or to support the people who will be responsible for it. An increase in costs of training and/or consulting is also common. There is also the need for greater communication between departments; the exchange of ideas should be promoted and sources of friction minimized. These frictions may become evident as early as during the implementation phase for the reasons outlined in the previous section; they may also prevail after the implementation phase, particularly when defining the causes of nonconformities (especially if they result from external audits, whether performed by customers, certifying bodies, or regulatory bodies). Another consequence is the increased sensitivity towards the food safety issues that will promote changes in the behavior of employees, not only internally but also when dealing with consumers. The organization will also have the tendency to be more organized but more bureaucratic, so it is very important to focus on creating simple, practical, and only relevant documentation.

### 3.4.3 Factors for success

There are several factors that may contribute to the successful implementation of the standard, as follows.

- *Top management commitment.* As mentioned in the previous section, it is only with a top management that is actively present and complying with the standard principles that achieving full success of the objectives planned in the standard will be possible.
- *Involvement of all stakeholders.* In order to elaborate documentation and to maintain the standard operating effectively, it is important to take into consideration all contributions and suggestions, especially from the official health authorities, business and consumers associations, workers, suppliers, and customers.
- *Contribution of service-providing companies.* The smaller the size of the organization, the more important is the support of consultants and their continuous availability. As for audit firms, they are very relevant partners for the enrichment and continuous improvement of the system. Technical support from companies providing services that are relevant to food safety, such as accredited laboratories, pest control, maintenance and repair of equipment, and transportation of food products, should not be dismissed.
- *Continuous improvement of processes, knowledge, and procedures.* The relevance of this aspect is visible in the standard itself, as it has a clause (8.5) dedicated to it. A well-implemented food safety system is a dynamic system which needs periodic reviews to adjust it to new requirements (legal, regulatory, or from suppliers), to adapt to new structural realities or system performance, and to eliminate the causes that lead to nonconformities.
- *Food Safety Team leader.* The choice person responsible for leading the process is dependent on the size/complexity of the organization and on the available team. The three most important characteristics are as follows.
  1. *Technical knowledge.* Even in the case of small enterprises that use consultants, the person responsible for the team must always have knowledge of the standard and the process. The greater the size and complexity of the organization and of the processes, the greater the need for someone with acquired knowledge and experience of the industry. The presence of others with relevant knowledge in the Food Safety Team to support the leader is also important.
  2. *Organization.* Good organizational ability of the leader becomes more important with reduced levels of support. The management system, particularly in the documentary aspect, requires someone with the ability to be methodical in maintaining the system updated. If the leader is able to delegate these activities to a collaborator, or if there is a consulting firm, this characteristic is not as important.
  3. *Interpersonal relationship.* Internal and external communication is one of the most focused requirements throughout the standard. In this sense, the ability of the leader to communicate interactively with all stakeholders is essential, regardless of the size or complexity of the organization.

## 3.5 Technical specification ISO/TS 22002-1

ISO 22000 describes how the organization shall establish, implement, and maintain a prerequisite program that can help control the hazards of food safety to which products may be exposed. As this standard was created in order to be applicable to any organization of the food chain, there are 10 points (see below) that organizations should consider while developing their prerequisite program; this list lacks specificity and guidance, however. Moreover, even in ISO 22004:2014 *Guidance on the application of ISO 22000:2005* no actual guidance is given for the implementation of Clause 7.2.

1. construction and layout of buildings and associated utilities;
2. layout of premises, including workspace and employee facilities;
3. supplies of air, water, energy, and other utilities;
4. supporting services, including waste and sewage disposal;
5. suitability of equipment and its accessibility for cleaning, maintenance, and preventive maintenance;
6. management of purchased materials, supplies, disposals, and handling of products;
7. measures for the prevention of cross-contamination;
8. cleaning and sanitizing;
9. pest control; and
10. personnel hygiene.

The ISO/TS 22002 consists of a series of technical specifications with the designation *Prerequisite program on food safety*, in which the requirements for the implementation of the prerequisite program for each of the following areas are specified:

• Part 1: Food manufacturing;
• Part 2: Catering;
• Part 3: Farming; and
• Part 4: Food packaging manufacturing

ISO/TS 22002-1 is intended for all organizations whose activity is manufacturing, regardless of their size and complexity. It adds 5 new points to the 10 points presented in Clause 7.2 of ISO 22000:2005, as they are considered relevant for this step of the food chain. The additional points are:

1. rework;
2. product recall procedures;
3. warehousing;
4. product information and consumer awareness; and
5. food defense, biovigilance, and bioterrorism.

This technical specification defines in detail the requirements that must be met in the various points of the prerequisite program, facilitating not only the implementation but also the verification of compliance during audits. More information on this technical specification is presented in Sections 4.5.2 and 5.3.

## CHAPTER 4
# Food safety management system EN ISO 22000:2005

## 4.1 Introduction (Clauses 1–3)

**Figure 4.1** Keywords from Section 4.1.

This international standard defines requirements in terms of food safety applied to organizations in the food chain or organizations that support it (see Fig. 4.1 for relevant keywords). In 2004, 54 experts were registered as members of the working group responsible for the development of ISO 22000:2005, and the decision to publish it was taken unanimously by the participating countries of ISO/TC 34/SC 17.[1]

Despite the high number of participants of this committee, any change to the standard can only be approved when 75% of its voting members are in agreement (ISO 2005a). This fact provides this international standard with a wide range of inputs and, at the same time, stability and credibility. Although the requirements

[1] Information obtained from ISO/TC 34/SC 17 secretariat.

*Food Safety in the Seafood Industry: A Practical Guide for ISO 22000 and FSSC 22000 Implementation,*
First Edition. Nuno F. Soares, Cristina M. A. Martins and António A. Vicente.
© 2016 John Wiley & Sons, Ltd. Published 2016 by John Wiley & Sons, Ltd.

that organizations must comply with are only stated as of Clause 4, some of the most important concepts are introduced before. Those concepts can be fundamental to a successful implementation of the standard in an organization.

Throughout the food chain, several food safety hazards may occur. Collaboration between all parties involved in the food chain is therefore fundamental, providing confidence that, at each step, food safety requirements have been complied with.

The introduction of this standard presents four key elements that are present throughout the norm: interactive communication; system management; prerequisite programs; and HACCP principles. These are fundamental to guarantee food safety in every part of the food chain.

Interactive communication highlights the importance of interaction between each step within the food chain. An accurate and complete transmission of information between the food chain and external stakeholders will ensure, in a more efficient way, the identification and control of all relevant risks to food safety. It is therefore important that each organization understands its role and position in the food chain in order to request and provide the necessary information to guarantee food safety until final consumption.

The standard should be implemented as another tool of the overall organization and not as an isolated element, independent of management functions. The compatibility of this standard with other existing standards, particularly ISO 9001, allows its adaptation and integration when organizations are implementing other management systems.

Hazard analysis is essential in the implementation of an effective food safety management system; this standard integrates the HACCP principles and application steps developed by the *Codex Alimentarius*. In addition, it associates HACCP with prerequisite programs. This combination helps to organize the necessary knowledge to establish an effective range of control measures.

In the scope of the standard it is stated that 'this International Standard specifies requirements for a food safety management system where an organization in the food chain needs to demonstrate its ability to control food safety hazards in order to ensure that food is safe at the time of human consumption.' The scope is generic in terms of the type of organization that can implement it; 'It is applicable to all organizations, regardless of size, which are involved in any aspect of the food chain…' but specific for the type of food safety hazards that need to be controlled. However, organizations may find it useful to use the same approach to respond to other situations nonspecific to food safety.[2] Furthermore, the standard also mentions the possibility of organizations using external resources to meet the requirements

---

[2] Although the standard clearly states that it 'is intended to address aspects of food safety concerns only,' several of the procedures presented can be used in other aspects. The procedures established for employee training or the definition of responsibilities and authorities or communication with customers (e.g., enquiries, complaints) are some examples of situations in which the organization may use the same approach, and even the same documents/records, to address issues beyond food safety.

or, in the case of a small and/or less-developed organizations, the possibility of implementing a combination of control measures established externally.

ISO 22004:2014 clarifies that organizations may appeal to external assistance (e.g., generic guidelines or models, individuals, or organizations acting as consultants). However, the organization must guarantee that when using guidelines or models developed externally they are suitable, adjusted, and appropriate to the organization. This option seems particularly interesting for small and low-complex-activity businesses, since it will allow less use of resources in the implementation of the standard.

## 4.2  Food safety management system (Clause 4)

For keywords, please see Figure 4.2.

**Figure 4.2** Keywords from Section 4.2.

### 4.2.1  General requirements (Clause 4.1)

This clause provides general requirements that the organization must fulfill and materializes some of the principles mentioned in Section 4.1. From the communication point of view, the organizations are compelled to share information about food safety throughout the food chain and within the organization. Even with different objectives, they share the same purpose: deliver safe products at the time of consumption. The principle of continuous improvement is also present in the requirement, since it is stated that the system should be periodically evaluated and updated with the most recent information available.

The standard also requires the definition of a scope that 'shall specify the products or product categories, processes and production sites that are addressed by the food safety management system.' The scope is very important because it clearly identifies the boundaries within which the organization must comply with the requirements of this standard. When this standard is applied as part of the FSSC 22000 the scope has to comply with what is defined in the scheme, as described in Section 5.2.1.

The possibility to 'outsource any process that may affect end product conformity' is also foreseen. When this is the case, the organization must identify and document how those processes are controlled. The organization must guarantee the same level of control as if the process was performed in-house.

## 4.2.2 Documentation requirements (Clause 4.2)

Documentation control is one of the foundations on which an organization must support its operating activities and is clearly a key element in the success of any management system. This control allows the organization to keep its documentation constantly available and updated at the appropriate locations in order to be used or consulted whenever necessary. In the application guide of the standard (ISO 22004:2014) it is specified that the documentation can be supported in any kind of media. ISO 22000:2005 divides the documentation in three main groups:

1. statement of the food safety policy and its objectives;
2. procedures and records required by the system; and
3. support documents for development, implementation, and update of the standard.

The statement of policy and related objectives is a particular type of document that the top management shall define, as described below in Section 4.3.2. The system procedures describe the activities that implement food safety and document the actions that need to be developed and respective responsibilities. The records provide evidence that the procedures established in the FSMS are implemented as described and in accordance with the ISO 22000 requirements. Both result directly from the need to comply implicitly or explicitly with the requirements that are presented throughout the scheme.

Support documentation consists of any document that, although not required or explicitly mentioned in the standard, is necessary for its development, implementation, and update (e.g., legislation, good practice guides, fact sheets).

The complexity and amount of documentation required varies depending on the dimension and complexity of the organization. However, there is documentation that must be present independently of these constraints (e.g., policy and objectives, prerequisite programs (PRPs), operational prerequisite programs (OPRPs), HACCP plans). Support documentation and the number of records are, on the other hand, examples of documentation that vary more from one organization to another.

ISO 22004:2014 divides the documentation into three main groups (Fig. 4.3) and presents for each one a comprehensive list of the documents necessary to comply with ISO 22000:2005 requirements. Appendix 1 contains the list of documents mentioned in the guide for each of the three groups. Activities that can be outsourced (e.g., laboratory analyses, transportation, storage, and pest control) should also be documented as part of the system.

**Figure 4.3** The three main groups in which documentation is divided, according to ISO 22004:2014.

The information required for implementation, maintenance, and update of the standard, particularly in bigger or more complex companies, can produce a large number of documents, leading to greater complexity and even discouraging management and other workers. There is therefore a tendency for the increasingly widespread use of computer support to provide and file documentation. This option can be very helpful because it allows a faster update and distribution of new documentation, while providing practical and automatic evidence of these activities. It also facilitates the internal/external communication. The use of dedicated software incurs costs that cannot always be supported by organizations. However, there are some tools which are available for free that can facilitate the implementation of FSMS, making document management a more user-friendly and less time-consuming process (see Box 4.1; Figs 4.4, 4.5). However, this alternative requires the application of measures to ensure protection, recovery, and retention of documents and information. The use of back-ups, databases, and antivirus tools and the definition of passwords and access restrictions are important to allow the creation of conditions for safe use of computer storage media.

After having identified the documentation required to be part of the FSMS (Clause 4.2.1), the following two clauses define the means to control it: Clause 4.2.2 and Clause 4.2.3.

Table 4.1 lists the requirements defined by the standard as mandatory to be included in the documented procedure for the control of documents and examples of how to implement them. A record is considered a special type of document with specific requirements for its control. Records are essential as evidence of the performance of the food safety management system. In order to attain this objective, records must be complete, legible, clearly identifiable and easily retrievable,

**Box 4.1**  Online surveys

It is possible to find online tools to develop surveys that can be used to reduce the amount of documents, while also providing easier ways to access and analyze the information obtained. Two examples where these tools can be used with great advantage are when the organization uses surveys to obtain information from suppliers and (adapting a survey) to make internal audits or checklists.

In the first case (Fig. 4.4), an online survey not only benefits the organization (it reduces the paper work, the time to prepare the survey and to reach the supplier, and is less time/paper consuming) but also benefits the supplier who has an easier and quicker way to comply with the customer requisite. In the second example (Fig. 4.5), the adaptation of a survey to complete audits or other checklists is hugely convenience, especially when a tablet computer with internet access can be used.

In both cases, the information obtained can be read by anyone granted access and at any location with access to the internet. The information is commonly saved in a spreadsheet or similar tool, enabling easy statistical evaluation of the results.

Box 4.2 explains how to use QR codes to provide access to food safety information but also as an example of using available online tools to minimize bureaucracy and increase interactive communication.

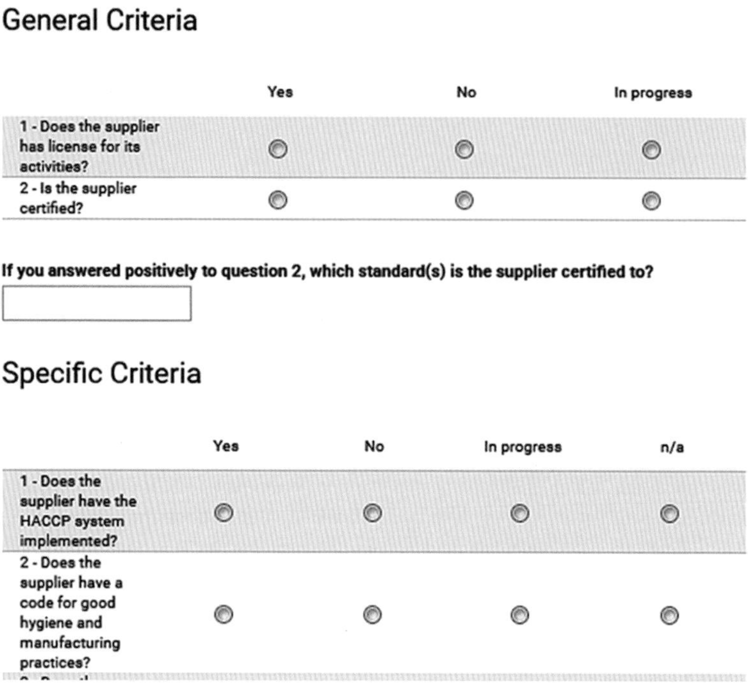

## General Criteria

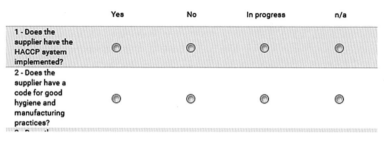

|  | Yes | No | In progress |
|---|---|---|---|
| 1 - Does the supplier has license for its activities? | ○ | ○ | ○ |
| 2 - Is the supplier certified? | ○ | ○ | ○ |

If you answered positively to question 2, which standard(s) is the supplier certified to?

## Specific Criteria

|  | Yes | No | In progress | n/a |
|---|---|---|---|---|
| 1 - Does the supplier have the HACCP system implemented? | ○ | ○ | ○ | ○ |
| 2 - Does the supplier have a code for good hygiene and manufacturing practices? | ○ | ○ | ○ | ○ |

**Figure 4.4** Example of an online survey (partial).

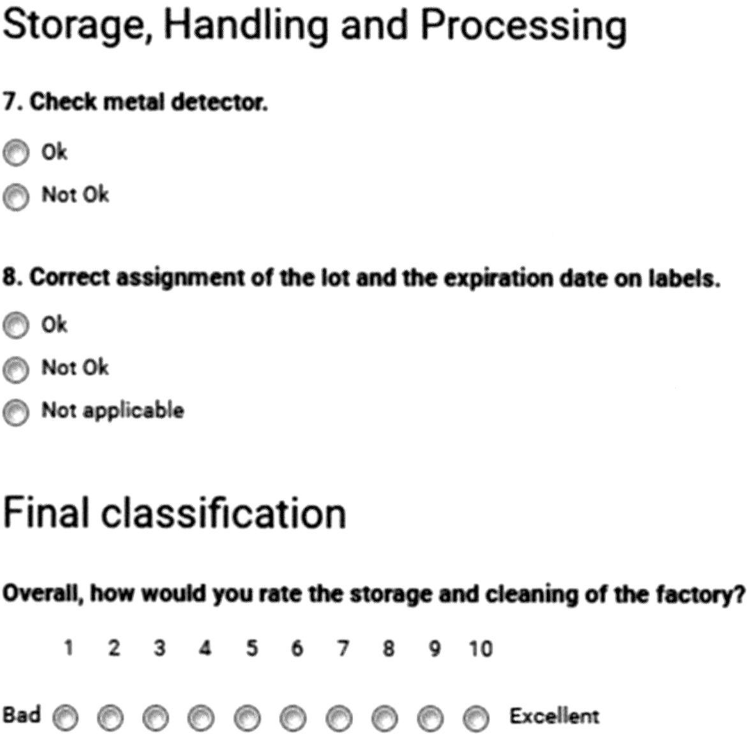

# Storage, Handling and Processing

### 7. Check metal detector.

○ Ok

○ Not Ok

### 8. Correct assignment of the lot and the expiration date on labels.

○ Ok

○ Not Ok

○ Not applicable

# Final classification

### Overall, how would you rate the storage and cleaning of the factory?

1   2   3   4   5   6   7   8   9   10

Bad ○ ○ ○ ○ ○ ○ ○ ○ ○ ○ Excellent

**Figure 4.5** Example of an online check (partial).

**Table 4.1** Procedure requirements and examples of methods of implemention

| Procedure requirements | Examples |
| --- | --- |
| Approve documents before use | Identify the designated personnel that have the knowledge and authority to verify adequacy and approve documentation. It is common that the food safety team leader (FSTL) or top management are responsible for this function, since it requires global knowledge of the standard, food safety, and organization processes. |
| Identify the need for a new version and update of documents | Suggestions for improving the documentation should be encouraged, although it is common that approval remains the responsibility of the FSTL or top management. A maximum period of time to verify that the document remains adequate can also be stated. |
| Identify the modifications made in the documents and the current version | Describe the methods of documenting each modification and identify the newest version. |
| Ensure that the latest version is available at the required location | Identify the location(s) where each document should be used and by whom, so that where to replace a document at each new version is clearly known. The process of changing the old documents for the new documents is described. |
| Ensure that the documents are identified and legible | A description of how each type of document is identified must be made. It is common to establish a 'template' for each document of the system, including information such as the document name, the version number, and the person responsible for development and approval. |
| Ensure the identification of the external documentation and the control of its distribution | Identify the process/responsibility to obtain/seek external information. According to each type of information it is important to define the periodicity of this research, ensuring that a routine is created within the organization. |
| Ensure that obsolete documents are identified and not used | Describe the procedures for the substitution of obsolete documents from the locations/personnel of use and define how to identify them clearly to avoid using them by mistake. |

have a defined and documented retention time, and be stored in safe locations which are protected from deterioration.

Table 4.2 lists the instructions that the standard provides to control records and examples of how they can be described in a procedure. To better understand how to control documents and records according to the standard requirements, an example of the contents of a record that can be created for that purpose is as follows.

- *Designation and/or code*: Clearly identify the document.
- *Revision*: Identifies the current version that is being used. If the organization wants to have documentation in paper support, it should consider the size of the documents that are codified under the same version since any change requires the alteration of the version for the entire document and a new print.
- *Date of approval*: Identifies the date of approval of the latest version.

**Table 4.2** Instructions to control records and examples of how can they be described in a procedure

| Instructions | Examples |
| --- | --- |
| Identification | It must be described how each type of record is identified. It is common to establish a 'template' for each record of the system including information such as the record name, its version, and person responsible for development and approval. |
| Storage | Identify the location of and the person responsible for the records storage. |
| Protection | The method implemented to guarantee the security and confidentiality of records during the retention time must be described. |
| Retrieval | Describe the process of removal of complete or obsolete records from their location of use. |
| Retention time | Establish the retention time considering the expected usage of the products and shelf life along the food chain. |
| Disposition | Identify the personnel involved, locations, and methods of discarding records. |

- *Responsible for approval*: Identifies the person or persons responsible for approving the documents and records.
- *Distribution*: Identifies the place where documents/records are stored and used. In the case of records, information regarding the person(s) (or their function) responsible for its completion can be included.
- *Changes*: Identifies the modifications that were made in the old version.
- *Retention time*: Identifies the records storage time until their destruction.
- *Disposal*: Identifies the method to destroy the records.

## 4.3  Management responsibility (Clause 5)

For relevant keywords, please see Figure 4.6.

**Figure 4.6** Keywords from Section 4.3.

## 4.3.1 Management commitment (Clause 5.1)

Top management is defined according to the standard ISO 9000:2005 as a group constituted by the person or group of people who directs and controls an organization at the highest level, that is, those who occupy higher hierarchical positions (administration/management/general direction) and therefore have the autonomy to make decisions regarding the availability of resources necessary to achieve food safety (both in terms of material resources and human resources) (ISO 2005b). Workers who exercise functions of direction or department management may be considered top management in case they enjoy that autonomy.

The standard identifies methods of how top management should demonstrate its commitment to the development, implementation, and update of the food safety system (see Table 4.3). However, the documented evidence is not an absolute guarantee of compliance. The real commitment comes from the way in which top management is involved in the development of the system, in which is itself an example of compliance with established procedures while embodying the food safety system and its policy (see Section 5.2). Without a proactive approach of the top management, all efforts made by the rest of the organization may be insignificant and eventually disappear with time. It is essential that top management create a culture where employees know they are valued and recognized as much as for fulfilling the requirements/objectives of food safety as for accomplishing other objectives of the organization (e.g., commercial objectives).

**Table 4.3** Evidence of the top management's commitment

| Requirements | Examples |
| --- | --- |
| Show that food safety is supported by the business objectives. | Evidence that the strategic vision of the organization and its corporate objectives incorporate and respect the principles of food safety. |
| Communicate the importance of meeting the requirements of ISO 22000, any statutory and regulatory requirements, and customer food safety requirements. | Minutes of meetings and records of the employees' training about statutory and regulatory requirements and customer food safety requirements. |
| Establish the food safety policy. | Evidence of knowledge about the food safety policy and objectives (see Section 4.3.2). |
| Conduct management reviews. | Minutes of the review meetings in accordance with Section 4.3.8 where top management participation (collaborating in the FSMS performance evaluation and its continuous improvement) is demonstrated. |
| Ensure availability of resources. | Guarantee that the food safety system is operational and not compromised by lack of human and material resources (Section 4.4.). |

## 4.3.2 Food safety policy (Clause 5.2)

As referred to in ISO 22004:2014, the food safety policy is defined by the top management as 'the basis of any organization's food safety management system.' The food safety policy is defined in Clause 3.4 of ISO 22000:2005 as 'overall intentions and direction of an organization related to food safety.' This standard identifies six requirements for the policy that the top management must enforce.

1. *It is appropriate to the role of the organization in the food chain*: the activity that the organization carries out, its complexity, and its relative location in the food chain should be considered in the policy. It is easy understandable that the objectives and policies from a primary production, food retail, or manufacturing organization are different.

2. *It demonstrates the organization's commitment to comply with statutory, regulatory, and customer requirements.*

3. *It is implemented, communicated and maintained at all levels of the organization*: management must use means to communicate the food safety policy to all levels of the organization, such as training or printing and displaying the information in places where it can be seen by all employees. Even if the policy is not too extensive (as recommended) it should be communicated in order to identify key points (particularly those designed to meet these requirements) that can be easily recognized and retained by personnel. Top management must guarantee that the policy is comprehended and adopted.

4. *It should be periodically reviewed to ensure its suitability*: the review is usually performed at least once a year or at a time that the management review is made (Section 4.3.8).

5. *It highlights the importance of communication* in order to guarantee food safety.

6. *It is supported by measurable objectives*: in order to define the food safety objectives, the organization must take into consideration the fact that they must be rigorous but achievable. Another important aspect is that they have to be easy to monitor and regularly evaluated. If during the period established for the achievement of objectives (this period should not be longer than the period of policy revision) it is clear that an objective will not be fulfilled, its immediate revision should be considered. It is common to define objectives regarding the number of recalls/withdrawals, occurrence of foreign bodies, number of complaints, analytical plan and internal audit results, number of training activities, and effectiveness of corrective actions.

## 4.3.3 Food Safety Management System planning (Clause 5.3)

The standard establishes the obligation of top management to ensure that the planning of the FSMS is carried out in order to fulfill the general requirements of Clause 4.1 and the food safety objectives. Top management should also ensure the constant integrity of the FSMS whenever updates are implemented.

It is important to emphasize that planning is fundamental to the success of the food safety management system. In fact, the decision to implement it, independently of using external help, should be taken after analyzing the current situation of the organization in terms of food safety, assessing knowledge of internal resources on the subject, and according to the size and complexity of the organization and the objectives established for the implementation of the FSMS.

No guidance is given for the application of this clause in ISO/TS 2004:2014; there is therefore no formal definition for the content of these plans. An example of application is the establishment of a plan in order to achieve (new) food safety objectives. Another example is the definition of a plan to change a production line or to develop new products. This plan may include:

- identification of the suggested modifications;
- definition of the responsibility for approving/making the alterations;
- definition of the responsibility for analyzing the impact of these modifications on the FSMS;
- identification of the necessary corrections to the system and the person who supervises and approves the adjustments; and
- assessment of the integrity of the system.

### 4.3.4 Responsibility and authority (Clause 5.4)

Throughout the standard the need to define responsibilities and authorities for the implementation of certain activities/tasks are referred to, as listed in Table 4.4. The identification of these responsibilities and authorities does not imply that top management cannot establish others that may be necessary for the operation and maintenance of the FSMS. It may also be advantageous, particularly for organizations where a management system is not implemented, to use this approach for other purposes (not related to food safety), as mentioned previously in Section 4.1.

For this requirement organizations may develop the organization's chart and job descriptions for example, where the following features should be defined:

1. essential/desirable skills;
2. essential/desirable education or training;
3. responsibilities;
4. authorities; and
5. person to whom problems related to FSMS should be reported.

### 4.3.5 Food Safety Team leader (Clause 5.5)

The Food Safety Team leader is the central element of the Food Safety Team (Section 4.5.3) and is elected by the top management. The team leader is critical to the success of the FSMS and, although this role is commonly attributed to the Quality Manager of the organization, it is fundamental to select or prepare someone who, in addition to technical skills, possesses other capabilities of equal or greater importance in order to take food safety to the plant floor and make it part of the company's culture. Examples of such capabilities are organization, leadership, communication, strong interpersonal skills, and the ability to inspire and motivate personnel.

ISO 22004:2014 suggests that the team leader should be a member of the organization with an understanding of the specificity of its hazards and an extensive

**Table 4.4** Identification of the clauses describing the need to define responsibilities and authorities

| ISO 22000:2005 Clauses | Transcriptions |
| --- | --- |
| 5.4 Responsibility and authority | All personnel shall have responsibility to report problems with the food safety management system … Designated personnel shall have defined responsibility and authority to initiate and record actions. |
| 5.5 Food safety team leader | Top management shall appoint a food safety team leader who, irrespective of other responsibilities, shall have the responsibility and authority … |
| 5.6.1 External communication | Designated personnel shall have defined responsibility and authority to communicate externally … |
| 6.2 Human resources | Where the assistance of external experts is required for the development … records of agreement or contracts defining the responsibility and authority of external experts shall be available. |
| 7.5 Establishing the operational prerequisite program (PRPs) | The operational PRPs shall be documented and shall include … responsibilities and authorities … |
| 7.6.1 HACCP plan | The HACCP plan shall be documented and shall include … responsibilities and authorities … |
| 7.6.4 System for the monitoring of critical control points | … responsibility and authority related to monitoring and evaluation of monitoring results … |
| 7.8 Verification planning | Verification planning shall define the purpose, methods, frequencies, and responsibilities for the verification activities. |
| 7.10.1 Corrections | All corrections shall be approved by the responsible person(s) … |
| 7.10.2 Corrective actions | Data derived from the monitoring of operational PRPs and CCPs shall be evaluated by designated person(s) with sufficient knowledge … and authority … to initiate corrective actions. |
| 7.10.4 Withdrawals | … top management shall appoint personnel having the authority to initiate a withdrawal and personnel responsible for executing the withdrawal … |
| 8.4.1 Internal audit | The responsibilities and requirements for planning and conducting audits, and for reporting results and maintaining records, shall be defined in a documented procedure. |

knowledge of hygiene, food safety management, and application of the HACCP principles.[3] The use of external resources is not rejected and could be considered, particularly in small or low-complex organizations. When this happens, it is recommended that a person from both the Food Safety Team and the organization who is able to maintain a frequent communication with the team leader is identified.

Although the use of external resources to acquire technical knowledge is possible, the task of supervising and motivating is difficult to someone from outside

---

[3] This has been considered more important throughout the years since the previous version of ISO/TS 22004:2014 (ISO/TS 22004:2005) recommended a basic knowledge of hygiene management and HACCP principles application.

the organization. However, external resources have the advantage of not being conditioned by other responsibilities within the organization. In fact, the importance of avoiding conflicts of interest whenever the team leader assumes other functions in the company, is highlighted in the standard. The standard identifies that the team leader shall at least have the responsibility and authority to:

- organize and manage the work of the Food Safety Team;
- ensure that all elements of the Food Safety Team have the training and relevant knowledge;
- ensure that the FSMS is established, implemented, maintained, and updated; and
- report the effectiveness and suitability of the FSMS to the top management.

Other responsibilities that are commonly assigned to the Food Safety Team leader include:

- communicating with external parties on matters related to food safety;
- planning the management review meeting (Section 4.3.8);
- monitoring audits to the organization and planning and monitoring audits to suppliers;
- coordinating training activities in the context of the FSMS; and
- providing advice to the top management regarding food and safety issues.

## 4.3.6 Communication (Clause 5.6)

As mentioned in Section 4.1, the standard considers interactive communication a key element. Communication is also referred in the general requirements (Section 4.2.1), highlighting the importance of maintaining it throughout the food chain and throughout the organization. The standard identifies specific requirements for both external and internal communication.

### External communication (Clause 5.6.1)

With the increasing dimension and complexity of the food chain, food products go through several stages, organizations, and even countries before reaching the consumer. The ISO 22000:2005 defines food safety as a concept which implies that food cannot 'cause harm to the consumer when it is prepared and/or eaten according to its intended use.' To achieve this objective all elements of the food chain must not only take responsibility for ensuring food safety during their processes, but also obtain and transmit to external organizations all relevant information to ensure the safety of products until the moment of consumption.

The standard emphasizes the importance of communication regarding food safety between organizations and with statutory and regulatory authorities and customers. Records that evidence this communication and documents defining the requirements of the statutory and regulatory authorities must be maintained.

Effective external communication implies that the organization provides and obtains relevant information on food safety without ambiguity or possibility of misinterpretation. This can be a challenge to companies operating in several markets due to language issues. Every time communication is made between two people or organizations that do not share the same native language, actions should be taken in order to ensure mutual understanding. Examples about the kind of information that may be required or made available externally are described in the following sections.

*Suppliers and contractors*   When organizations define the suppliers and contractors that have a greater impact on food safety, it is common to emphasize the importance of communication with suppliers of raw material, goods, packaging, and hygiene/cleaning materials. However, there are other organizations whose services also need to be evaluated regarding their impact on food safety (e.g., pest control organizations, maintenance services, or waste collection services). Examples of information that can be exchanged with suppliers and contractors are provided below.

*Suppliers of raw material and goods*:
- Agreement on the food safety level required, such as defining microbiological/physicochemical criteria or other special requirements to be verified at the moment of reception.
- Information about suppliers, such as evidence of compliance with statutory and regulatory requirements, implementation of a food safety management system, or other certifications.
- Technical information about the supplied products, including the information referred to in Section 4.5.3 and the identification of the need to control any particular hazards.
- Results from controls carried out on supplied products at the moment of reception or during processing and from laboratory analyses. Communication of customer complaints. Identification of causes and measures taken by suppliers for the reported nonconformities.
- Information relative to changes in product specifications or to the update of the technical information. Suppliers should notify their customers whenever the need to retain or withdraw products arises.
- Results of audits carried out on suppliers.

*Suppliers of packaging, hygiene and cleaning materials*:
- Information about suppliers, such as evidence of compliance with statutory and regulatory requirements, implementation of a food safety management system, or other certifications.
- Technical information regarding the cleaning and hygiene materials, which should include expected use and evidence of their suitability to the purpose for which they are designed.
- Technical information of packaging materials that includes evidence of their suitability for use in food products and the accomplishment of specific regulatory requirements, including established migration limits. The level of information and requirements established to ingredients should also be applied to direct food contact packaging.

*Contractors*:

- Contract or an equivalent document that identifies the service to be provided, its duration, and/or periodicity and responsibilities assigned to the service provider.
- Information and training on food safety requirements that have to be respected by employees from the services company when attending the organization's facilities (e.g, pest control, maintenance).
- Information from the contractors about detected occurrences that may affect the safety of food products, even if they are not defined in their responsibilities.

Establishing criteria for the assessment of suppliers and evaluating their degree of compliance can be a method of monitoring their performance and identifying those who need to improve, or even be replaced, if incapable of fulfilling the requirements. This subject is addressed in more detail in Section 4.5.1, Prerequisite 6.

*Customers or consumers*    The standard introduces several examples of relevant information that can be exchanged with customers or consumers in order to ensure that all appropriate knowledge to guarantee food safety is available throughout the food chain.

1. *Product information*: For customers, the information is generally organized in data sheets of products which should include, among other information, statements regarding the intended use, specific storage requirements, and shelf life. Section 4.5.3 ('Product characteristics') provides a list of finished product characteristics that shall be documented and that could also be considered when defining the information to be transmitted. Labels are the most important vehicles of information about food safety, as part of that information has to be included in the label by legal obligation. However, organizations may resort to other tools to communicate with customers, such as that presented in Box 4.2. In fact, product information/consumer awareness is considered a prerequisite according to ISO/TS 22002-1:2009 (Section 4.5.1, Prerequisite 14).

2. *Enquiries and customer feedback*: customers and consumers are a very important source of information that the organization should promote and use to assess its performance in subjects relating to food safety. Replies to inquiries and complaints from customers and consumers should be analyzed carefully and used as input to improve the FSMS (Section 4.6.5). A complaint management procedure should be implemented and the responsibility to gather and transmit the information to the organization, make the analysis of causes, and define corrective actions and corrections (Section 4.5.10) should be assigned.

3. *Contracts or order handling*: the definition of contracts may be the best way to formalize mutual acceptance levels of food safety. This definition is especially important in subjects where there is no legislation or when it is intended to set more stringent limits.

**Box 4.2**   Food safety information at the point of use

The use of IT to share information related to food safety can have a huge impact on customers/consumers. In fact, this must be part of a widespread change in the mindset of organizations, particularly in its top management and/or food safety manager. Food safety information should not be limited to the fielding cabinet or hard drive, but should be available where it is most necessary: at the moment of use or consumption.

Quick response (QR) codes are two-dimensional barcodes that have a greater storage capacity than the standard linear barcodes that were first used in automotive industry in 1994. QR codes can be read by image reader, smartphone, or software applications, and are popular for their large storage capacity and versatility of the information contained (text, URL, location). When a uniform resource locator (URL) is placed inside a QR code, the user can be redirected to relevant information related to food safety, reducing the mishandling and misuse of the product. It is the most efficient and easily updated way to share information about food safety directly with consumers and also to obtain feedback if, for example, a small query is also available.

Since 2012, GS1 (a global organization that manages barcode standards to guarantee that they can be scanned anywhere in the world) introduced the GS1 QR code with the purpose of sharing extended packaging information. Logistics data that are contained in the GS1 QR code (e.g., lot, expiration date, Global Trade Item number or GTIN) can be read by any organization in the global market. Since 2011, GS1 and the Open Mobile Alliance (OMA), recognized for providing open specifications for the creation of services that operate across all geographical borders, have been working together to develop universal code specifications.

*Statutory and regulatory authorities and other organizations*   The standard identifies the importance of establishing channels of communication with the statutory and regulatory authorities, as well as with any other organization relevant to an efficient and up-to-date FSMS. Statutory and regulatory authorities are very important, not only as a source of information about legislation, but also to give assistance to its application. It is also common for these authorities to produce reports of their activities and publish notifications when the activity of an organization or the commercialization of a product is suspended due to food safety issues.

## Internal communication (Clause 5.6.2)

Internal communication is a fundamental tool to ensure the accomplishment of the food safety principles within the organization. Only the use of a holistic approach to communication allows certain behavior to be permanently modified. As mentioned in Section 4.3.1, one of the most effective methods is to 'communicate by example,' especially when it comes from the top management. At the

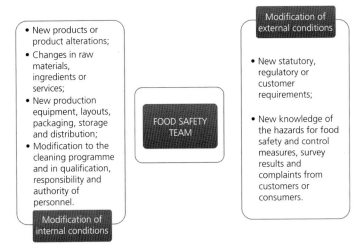

**Figure 4.7** Modifications that may compromise food safety.

same time the organization should create the means for all employees to access relevant information that will allow them to ensure the safety of food products. The top management and the Food Safety Team must create an internal communication dynamic on issues related to food safety. Besides the use of training sessions or public meetings, the use of audio-visual/electronic supports (e.g., internal television, website, intranet, newsletters) or even conventional methods such as information boards, signs, or slogans are all recommended.

In this clause of the standard, particular emphasis is placed on the obligation of informing the Food Safety Team of any change that may compromise food safety. Such changes can be grouped into two major groups, as shown in Figure 4.7. This information should be used in the FSMS update and included in a management review (Section 4.3.8).

## 4.3.7 Emergency preparedness and response (Clause 5.7)

The standard states that the top management should identify the emergencies or accidents likely to occur (considering their role and position in the food chain, geographical location, social stability, and country's politics) that may have an impact on food safety, and the relevant actions to eliminate those negative consequences. This procedure may contain information such as:

1. identification of the emergency or accident;
2. measures to contain or reverse the cause of the event;
3. procedure to identify the affected product;
4. alternative processes that could maintain the safety of affected products;
5. the process to evaluate the safety of affected products; and
6. definition of the responsibility for performing each of the activities above and to communicate with relevant stakeholders.

**Table 4.5** Description of procedures for dealing with emergency situations

| Task no. | Example 1 | Example 2 |
|---|---|---|
| 1 | Energy failure (refrigeration system stop) | Vehicle accident (distribution) |
| 2 | Trigger the generator and close or limit access to refrigerated or frozen storages. | Identify the actions to take depending on the estimated time that the vehicle will be immobilized. When the safety of the product is at risk, define procedures to send another vehicle in order to collect it or continue distribution. |
| 3 | Complete the record which identifies the occurrence and the affected product and place it next to the product or storage. | Complete the record which identifies the occurrence and the affected product. Define the procedure to identify and store the product returned to the organization. |
| 4 | Move the product to another defined place where safe conditions can be maintained. | Identify alternative storage facilities when the return of the product is not possible. Hire a company to continue the distribution or collection of the product. |
| 5 | Establish the periodicity to check if the product still remains at a temperature that guarantees its safety. Define actions to take when that does not happen. | When the safety of the product is questionable, it must return to the organization. Assessment may consist of microbiological or physical/chemical analysis, sensory evaluation by trained personnel, or by authorization of statutory and regulatory authorities. |
| 6 | Name the functions/personnel responsible for each of the above tasks and communicate with stakeholders. | |

The interpretive guide ISO 22004:2014 identifies natural disasters, environmental accidents, or bioterrorism[4] as examples of emergency situations. The importance of using test exercises to periodically verify the adequacy of the procedure and the organization's response to a particular situation is also emphasized. One common exercise that organizations should carry out regularly (at least annually) is a product withdrawal/recall (Section 4.5.1, Prerequisite 12 and Section 4.5.10). This process is particularly significant because it can be required whenever any of the emergency situations mentioned above occurs and the affected product reaches the market. It is also a test to traceability (Section 4.5.9).

Table 4.5 lists examples of how each of the six points highlighted above can be described in the procedure in case of energy failure and vehicle accidents.

---

[4] This topic is considered a prerequisite in ISO/TS 22002-1:2009 (Clause 18) and is developed in Section 4.5.1 of this book.

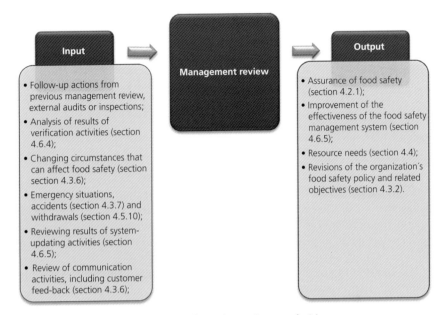

**Figure 4.8** Management review input information and output decisions.

## 4.3.8 Management review (Clause 5.8)

Management review is one of the procedures that top management uses to evaluate and ensure the continual improvement of the FSMS (Section 4.6.5). The standard establishes that top management must define a frequency for the review of the system to ensure its continuous suitability, adequacy, and effectiveness. No guidance is given for the maximum or minimum time between management reviews; however, it is not common for this period to be longer than a year (in order to support and promote continuous improvement and assess the food safety objectives and the adequacy of the food safety policy).

The standard identifies the information that is mandatory for discussion during management review and the outputs (decisions) of the meeting (Fig. 4.8). The success of the review is highly dependent on preparatory work, usually performed by the Food Safety Team, to gather relevant information and structure it in a manner that can be easily apprehended and compared with the food safety objectives (Section 4.3.2).

## 4.4 Resource management (Clause 6)

The nature of the organization's activities and improvement cycles determine the required resources and how often they have to be reviewed (Fig. 4.9 provides topic keywords). The quantity and/or nature of required resources are therefore dynamic within all organizations.

Infrastructures    Organization
# Human resources
Maintenance    Skills    Record    Standard
# Work environment
Food safety    Training

**Figure 4.9** Keywords from Section 4.4.

Although Resource Management is presented in ISO 22000:2005 separately from Management Responsibility, it is clear (Table 4.3) that it is one of the top management obligations. This means that top management must be committed to support the personnel elected to manage human resources, infrastructure, and work environment, in particular by providing the financial resources necessary to satisfy the standard requirements.

ISO 22004:2014 describes the importance of periodically monitoring, evaluating, optimizing, and reviewing the availability and suitability of resources.

### 4.4.1 Human resources (Clause 6.2)

The standard requires not only the Food Safety Team, but also all personnel that perform activities with an impact on food safety, to be competent and have appropriate education, training, skills, and experience.

An organization has three alternatives in meeting this requirement, either:

1. providing training to the personnel to achieve the skills defined as necessary for their activities;
2. hiring new personnel who already have the necessary skills; or
3. seeking the assistance of external experts with the necessary skills.

It is usual that organizations employ more than one of the alternatives listed above. In the particular case of using external experts, keeping records of contracts or agreements describing their responsibility and authority is necessary. This requirement is a particular aspect of a principle mentioned in the clause of external communication with contractors (Section 4.3.6).

The standard identifies the need for organizations to use training and effective communication when defining competences and when ensuring personnel are aware of the importance their actions have on food safety (Fig. 4.10).

The first step should be the identification of the necessary competences to perform any activity with an impact on food safety. Section 4.3.4 provides an example of a record in which this identification is defined. After the identification, training activities should be developed in order to ensure that the necessary skills are achieved. The standard gives particular emphasis to the importance of providing training to all those responsible for monitoring, making corrections, and taking corrective actions. It also requires the maintenance of records from training activities and the assessment of their efficacy. Box 4.3 provides an example

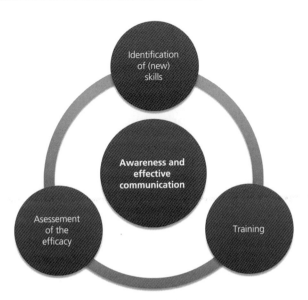

**Figure 4.10** Personnel development by training (cycle).

**Box 4.3**  'The hole keeps getting bigger'

Food safety incidents will always occur, but the important consideration is how they are handled. In 1972 Pillsbury rolled out the first HACCP system consisting of three principles: conduct a hazard analysis, determine critical control points (CCP), and establish monitoring procedures. Within a year, it was discovered that Pillsbury's dehydrated potato flakes were contaminated with small pieces of wire. Investigations revealed that the wire pieces came from a broken sifter screen, located just before the packaging machine. At headquarters we could not imagine how this could have happened, as that sifter had been designated as a CCP that required regular inspection. Sure enough, the sifter had indeed been inspected regularly and the inspector had first noted, 'Hole in sifter screen.' For several weeks he had diligently noted: 'Hole getting bigger.' He had done exactly what he had been told to do: inspect and record. It did not occur to him report this matter and there was no formal requirement for him to do so.

of an episode experienced by Dr William H. Sperber in the beginnings of HACCP, when the importance of communicating and reporting problems became clear.

The implementation guide of ISO 22000:2005 details information that should be included in the training record, such as: program content; name and qualifications of the trainer; final assessment of trainees; and establishment of the requirement for retraining. Other relevant information that can be included in that record is: names of the trainees; duration; and objectives of training.

If the organization does not have software for the management of human resources, it can easily develop a computer record where all personnel are identified (personnel number, name, age, and function) and any attended training recorded. Information about the training or just a code or link that allows access to the training record generated may also be included. Such a computer record can also be used to log and control the periodical medical examination required by the prerequisites (Section 4.5.2).

## 4.4.2 Infrastructure and work environment (Clauses 6.3 and 6.4)

Infrastructure and the work environment should provide appropriate conditions to produce safe and adequate food for its intended purpose (Box 4.4).

The construction and maintenance of infrastructures should:
- be appropriate to the nature of the available products;
- be preceded by a review of the statutory and regulatory requirements as well as relevant codes or standards for the sector; and
- take into account the organization's relative position in the food chain.

The top management is responsible for ensuring the necessary resources to allow the creation of the conditions described above. A similar approach should be taken by the organization regarding the establishment, management, and maintenance of the work environment. 'Work environment' refers to the set of conditions in which the work is performed, including physical, social, psychological, and environmental factors (e.g., temperature, humidity, composition/circulation of atmospheric air) (ISO 2005b).

**Box 4.4** 'We are entirely without a clue'

*This comprehensive story that Dr William H. Sperber shared shows the importance of prerequisites programs and the consequences of neglecting them.*

Around 1980 I was conducting a due diligence assessment in preparation for the potential acquisition of a frozen foods company. During the inspection of one plant I was taken to the area where frozen foods exited the blast freezer to be packaged, cased, and palletized for frozen storage and distribution. Walking through the area was difficult as the concrete floor was very slippery. I immediately noticed that the entire floor, more than 100 m², was liberally covered with rather small wood splinters. It was immediately obvious to me that the splinters came from stacks of wooden pallets, many in poor condition. I couldn't believe my eyes. Almost as a joke, I asked my tour guide, the facilities quality manager, 'Do you ever receive consumer complaints about wood splinters in your products?' 'Yes,' he replied. 'We get quite a few complaints like that.' 'Do you have any idea were the wood splinters are coming from?' I asked in exasperation. Incredibly, he answered, 'We've never been able to figure it out.' It was about this time that the US Food and Drug Administration had developed and promulgated Good Manufacturing Practices for food operations. The need for such educational training and operation requirements was obvious.

To this day GMPs are a major prerequisite program that supports the HACCP system of food safety management, along with Good Agricultural Practices, Good Distribution Practices, and Good Consumer Practices. The last of these is currently under development by several food safety leaders in the global food industry, with major educational support from many governmental and consumer advocacy organizations.

The construction and management of infrastructure and the work environment is also described in detail in the prerequisites presented in ISO/TS 22002-1, particularly in clauses 4 (Construction and layout of buildings) and 5 (Layout of premises and workspace), which are discussed in Section 4.5.2.

## 4.5 Planning and realization of safe products (Clause 7)

Figure 4.11 provides keywords relevant to this clause.

**Figure 4.11**  Word cloud for Section 4.5.

### 4.5.1 General (Clause 7.1)

Organizations should plan and develop processes that ensure the effectiveness of their activities in order to obtain safe products. The standard ISO 22000:2005 identifies the need to implement prerequisites programs, the HACCP plan, and the operational prerequisite programs to achieve this goal.

Prerequisite programs aim to control the general hygiene and ensure good manufacturing practices. They establish conditions for obtaining a hygienic environment throughout the food chain, without implying the control of specific hazards (ISO 2014). The HACCP plan manages the control measures identified by the organization as necessary to control the critical control points (ISO 2014). The operational prerequisite programs are designed to control the likelihood of introducing dangers to the food safety or their proliferation in products or work environments (ISO 2005c).

### 4.5.2 Prerequisite program (PRPs) (Clause 7.2)

An implemented prerequisite program helps to reduce the likelihood of introducing hazards in the product through microbiological, physical, or chemical contamination and the hazard levels in the product or work environment. These programs should be appropriate to the needs of the organization, including the size, type of operation, and the nature of the products that are produced or handled. The approval of the prerequisites should be the responsibility of the Food Safety Team and should be implemented throughout the entire production system.

The ISO/TS 22002 series was created with the purpose of assisting in the implementation of ISO 22000:2005. This standard was designed in order to be applied to any organization in the food chain, therefore presenting a general prerequisites program. This limitation was identified by a group of large companies in the food industry, including Kraft, Danone, Nestlé, Unilever, General Mills, and McDonalds who, in collaboration with the British Standards Institution (BSI) and other food manufacturing stakeholders, developed a prerequisite program on food manufacturing, namely PAS 220:2008. This specification was later adapted and replaced by ISO/TS 22002-1.

ISO/TS 22002-1:2009: Food manufacturing is applicable to all organizations that are involved in the manufacturing step of the food chain, regardless of their size or complexity. The standard specifies conditions for the establishment, implementation, and maintenance of prerequisite programs (PRPs) in order to control food safety hazards as specified in Clause 7.2 of ISO 22000:2005. This technical specification was the first (2009) of many with the purpose of reaching different sectors of the food industry (Fig. 4.12).

ISO/TS 22002-1 has 18 clauses, but it is only after the fourth (inclusive) that the standard defines specific requirements for prerequisites. The prerequisites are described in the following sections according to the structure of ISO/TS 22002-1:2009; Prerequisite 1 corresponds to Clause 4 of the technical specification, Prerequisite 2 corresponds to Clause 5, and so on.[5]

The following documents were also consulted to support the explanation and examples provided:
- *Codex Alimentarius* Commission, General Principles of Food Hygiene (CAC/RCP 1-1969 v.4 2003) (CAC 1969, 2003);

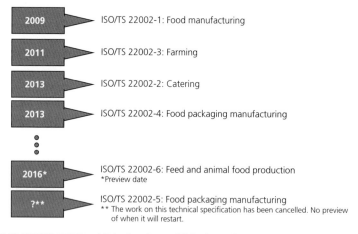

**Figure 4.12** ISO/TS 22002 published and unpublished specifications.

---

[5] A table listing all prerequisites is provided in Chapter 5 (Table 5.1), showing the correspondence between the clauses of ISO/TS 22002-1:2009 and the prerequisites presented in this chapter.

- Food and Drug Administration, 2013 Food Code (FDA 2013);
- Safe Quality Food Institute, General Guidance for Developing, Documenting, Implementing, Maintaining, and Auditing a SQF System – Module 11: Good Manufacturing Practices for Processing of Food Products (SQF 2014a);
- Clever *et al.* (2015) China – Peoples Republic of China's General Hygiene Regulation for Food Production (GB14881);
- Food and Drug Administration, Code of Federal Regulations – Title 21: Food and Drugs (FDA 2012b);
- Food and Drug Administration, Fish and Fishery – Products Hazards and Controls Guidance, Fourth Edition (FDA 2011); and
- *Codex Alimentarius* Commission, Code of Practice for Fish and Fishery Products (CAC/RCP 52-2003 rev. 2013) (CAC 2003).

## Prerequisite 1: Construction and layout of buildings

Buildings must provide adequate space to the nature of the operations that are carried out there; the flow of materials, products, and personnel must respect a certain logic and a physical separation between materials and waste should be kept in order to avoid cross-contamination by microorganisms.

The facilities of a food establishment should be located away from possible sources of contamination and the site should be cleaned to prevent the existence of objects that could facilitate pest infestation. Site boundaries shall be clearly identified and the access to the site must be controlled. The entries and parking areas must have a draining system to prevent standing water. Outside the building, the floor immediately in front of the doors and entries should be paved in order to minimize dust. At least once a year, the effectiveness of the adopted measures to control possible contaminants should be evaluated and reviewed if necessary. Box 4.5 presents examples for fish units in land and on a ship.

## Prerequisite 2: Layout of premises and workspace
### Internal design, layout, and traffic patterns

The facilities shall have a product flow pattern that is designed to prevent cross-contamination between the finished products and raw materials and to minimize delays (which can result in product quality loss or compromise its safety). This flow pattern should be respected and executed continuously so that there is total control of critical factors, such as temperature and time.

Facilities must also be designed so that there is a designated area for the entry and exit of personnel who manipulate food products, as well as a physical separation between the areas for raw materials and processed products. The technical specification states that a sufficient distance should be adopted[6] in order to minimize the risk of contamination between two materials.

---

[6] The standard does not ascertain a distance. This option should be used only when no other option is technically possible.

**Box 4.5** Construction and layout of buildings (example of fish unit in land and in ship)

To project a fish unit, the following physical and geographical factors of an appropriate location must be considered:

- size of the land: if it is appropriate to the current needs and future development;
- accessibility: by road and/or railways;
- water quality, energy, and waste removal/treatment services: should be appropriated and available throughout the year;
- waste removal: construction, design, location, and suitability of the space designed for that purpose; and
- pollution of adjacent areas – Evaluate the contamination of future facilities by air, through smoke, dust, ash or unpleasant odours present in the region.

**Particular case: Ships**

To project a ship it is important to be aware of certain aspects in order to minimize product contamination or deterioration.

- A good draining system must be in place to prevent standing water, which may cause the proliferation of microorganisms.
- Construction of interior walls must be avoided in order to facilitate cleaning and sterilization, and to prevent the accumulation of dirt.
- Harmful substances from the ship, including smoke, fuel oil, and water from the ship's hold, must not contaminate the fish.
- The containers for offal and waste material should be clearly identified and be made of a waterproof material.
- The entry of birds, insects, or other pests into the workplace should be prevented.
- Ships designed and equipped to preserve fishery products for more than 24 hours should have holds, tanks, and vessels to freeze or refrigerate the products, respecting the temperatures established for that purpose.

## Internal structures and fittings

Within the facility, all surfaces that are in contact with the product must be resistant to corrosion, made of a waterproof material, light colored, flat, and easily cleanable. The ceiling and overhead fixtures must be prepared to minimize the accumulation of dirt and the falling of particles. The internal walls must be easily cleanable and made of nontoxic and corrosion-proof materials.

The floor must be resistant to dropped products, water, and disinfectants and must be nonslip. The facility must have a water draining system that guarantees an appropriate flow. This system should include grids and/or removable drains to allow the easy cleaning of the facility. Corners between walls and floors must be designed to prevent accumulation of dirt.

Windows should be constructed to minimize the accumulation of dirt and must be protected with a mesh to prevent the entry of insects. Meshes must be removable and made of washable materials.

Doors must be flat, made of waterproof and washable materials, and guarantee the effective isolation between areas. They must always be closed when not in use. The doors and internal openings should be designed to minimize the entry of exterior materials and pests. It is advisable to use doors that close automatically (e.g., roll-up or swinging doors).

### Location of equipment
Equipment shall be constructed of removable or easily transportable components to allow its maintenance, cleaning, disinfection, and monitoring. It should be designed in order to minimize corners (prevent the accumulation of dirt).

### Laboratory facilities
Laboratory facilities should not have direct access to the production area and must be located and operated in a way that prevents contamination of food products. Their location shall take into account the level of risk that it might pose to the product. For example, if the laboratory manipulates pathogenic microorganisms, it must be located far from the production area.

### Temporary or mobile premises and vending machines
Vending machines shall be constructed in such a way as to avoid food contamination and pest harborage. When defining its location, the organization must take into account the risk of product contamination and consider reinforcing pest control (Prerequisite 9).

### Storage of food, packaging materials, ingredients, and nonfood chemicals
The facilities used for storage should protect the products from different sources of contamination (e.g., dust, waste, condensation drains). Storage areas shall be well ventilated and guarantee the ideal conditions of temperature and humidity defined for each food product. They should be designed to allow the separation of raw materials, work in progress, and finished products. The products shall be stored off the floor with easy access to allow the realization of inspection, cleaning, and pest control activities.

The facilities should have a specific area to keep the cleaning products, chemicals, and other hazardous substances. Access to these materials must be controlled in order to prevent their careless use which may constitute not only a risk to the health of personnel but also a risk to the product if, for example, excessive amounts of them are used. It is recommended that access to the storage is prevented or is limited to personnel with specific training.

The storage process should be appropriate to avoid crushing or breaking the product or packaging. The fact that a product can cause damage to other products by putting pressure on them should be considered and avoided. The personnel in charge of these operations should have relevant training, particularly in the use of forklifts or pallet jacks.

## Prerequisite 3: Utilities: air, water, energy
### General requirements

The sources of water, air, and energy shall be controlled in order to guarantee their quality and minimize the risk of product contamination.

The distribution routes for these utilities must be designed in order to avoid the risk of cross-contamination and be monitored to avoid water and air contamination.

### Water supply

The quality of the water should be in accordance with the needs of the process. When water is used as an ingredient, to wash food or food contact surfaces, or to manufacture ice or steam that comes into contact with food, it must comply with the chemical, physical, and microbiological requirements specified for potable water and the product in question.

The supply of potable water should be sufficient to meet the needs of the process and be used in all processes that are directly or indirectly in contact with food to avoid contamination. Nonpotable water must have a separate supply system (clearly identified) and the mixing with potable water prevented.

It is common to use chlorine in the treatment of water, especially when the facility has its own water supply system. When this is the case, a procedure must be put in place in order to control the amount of chlorine added to water. Depending on several circumstances (e.g., distance from the injection/addition point to the point of use, temperature, time) the residual chlorine content will vary and must be controlled according to appropriate requirements. Whenever possible, the use of potable water supplied by organizations monitored or controlled by legal authorities is recommended.

It is also recommended to identify all water supply points in the production area as well as in the production plan. This facilitates water quality monitoring (usually performed in alternate points) and the identification of points where the assessment is more critical, requiring more frequent and/or rigorous control.

### Boiler chemicals

When chemical additives are used in the production of steam, they must be authorized by a regulatory authority to ensure that the additive is approved for human consumption (approved as food additive or safe to use in water for human consumption). As described in Prerequisite 2 (Layout of premises and workspace), the chemicals should be stored in a separate and secure area when they are not being used.

### Air quality and ventilation

The air in direct contact with the product or product contact surfaces should not constitute a hazard to the safety of the product. In order to minimize the risk of contamination, the organization must control/monitor air quality (especially in areas where products are exposed) by filtration systems and by setting humidity and/or microbiological parameters.

The ventilation system should be sufficient to remove the excessive steam, smoke, and unpleasant odors and must be constructed to avoid the mechanical flow of air from contaminated to clean areas. These systems must be easily accessible for the purposes of maintenance (e.g., filter replacement).

In addition to assessing air quality against established requirements, it is also relevant to compare it with the outside air and to analyze trends over time. The assessment of the inside air quality when compared with the outside air quality may indicate whether the latter is a source of contamination and to what degree. Analysis of the results allows tendencies to be established and proactive action taken (when necessary) in corrective actions. It is recommended that the comparison of the obtained results be performed at the same time each year in order to minimize the influence of variations in outdoor air quality (resulting from natural climate changes experienced during different seasons) on the results.

### Compressed air and other gases

The compression systems of air and other possible gases that are present in production units shall be constructed and maintained in good condition in order to prevent leaks and the contamination of food products. Filtration systems should be located as close as possible to the point of use. The gases that are in contact with products must be approved for that purpose and be free from dust, oil, and water. It is not recommended to use oil in the compressor but, if there is no alternative, the oil used must be of food grade. The level of control of air or other gases depends on the type of product.

### Lighting

Adequate lighting shall be provided in all working areas to allow operation in a hygienic manner. Lighting can be natural or artificial and should be appropriate to the nature of the operation. The FDA (2013) gives an example of the level of light intensity in each operation:

- >108 lux at a distance of 75 cm (30 in) from the ground in refrigeration units and storage areas of dry products and other areas during the cleaning period;
- >215 lux at a distance of 75 cm (30 in) from the floor in bathrooms, areas to wash the hands, and in storage areas of equipment and utensils;
- >540 lux in production places where personnel use sharp utensils such as knives, slicers, and saws.

It is possible to find similar values in the Code of Hygienic Practice for Low-Acid and Acidified Low-Acid Canned Foods (CAC/RCP 23-1979; CAC 1979) which defines 540 lux at inspection points, 220 lux in working areas, and 110 lux for the remaining areas.

Light fixtures shall be protected to ensure that the product is not contaminated in case of breakage. The organization should be aware of any lights that are not easily accessible (e.g., inside equipment) and take measures to protect them.

## Prerequisite 4: Waste disposal
### General requirements

Waste management systems should be implemented to prevent the contamination of products or production areas and ensure its adequate elimination.

### Containers for waste and inedible or hazardous substances

Containers for waste help to prevent its accumulation in the production area and shall be: clearly identified for their intended purpose; located in a designated area; constructed of a waterproof and washable material; and closed unless they are being used continuously.

The location established for the containers shall be considered carefully since it implies: tighter control of pests at that location; special attention in the cleaning program; and deterioration of air quality.

The paper bin used for hand hygiene should be located near the place of washing and used appropriately to avoid contamination. Its contents shall be removed regularly, usually at the end of the day.

### Waste management and removal

Waste removal frequencies shall be established according to waste category and to avoid accumulation. ISO/TS 20002-1 states that waste must be removed at least daily. When an external organization is responsible for waste collection and destruction, it must be properly approved for it. Records of its service shall be kept for the period legally established or for the period of time relevant for traceability.

All materials that contain trademarks and are considered waste shall be disfigured or destroyed. This destruction must be accompanied by someone from the organization to ensure that these materials cannot be reused in a malicious way (the importance of organizations to be alert to these situations is reinforced in Prerequisite 15). The organization shall retain records describing the destruction.

### Drains and drainage

A water drainage system should be implemented throughout the facility allowing an appropriate flow. This system should include grids and/or removable drains, which must always be placed to avoid reflux of unpleasant odors, the entry of pests, and to facilitate the cleaning of the facility. The draining system should be installed to prevent the flow from a contaminated area to a clean area; when it is not possible to have two separate systems it should be ensured at the time of installation that water runs from a clean area to a contaminated area, and not the opposite. Drainage systems shall be properly identified in the installation plan.

## Prerequisite 5: Equipment suitability, cleaning, and maintenance
### General requirements

The materials in direct or indirect contact with food products should be appropriate and must not represent a risk to the health of consumers (e.g., migration of substances from materials to products) or even cause an unacceptable change in the organoleptic properties of the products. On the other hand, these same materials should not be affected by products or cleaning agents. All equipment shall have an instruction manual and certificates of conformity that prove compliance with statutory requirements.

## Hygienic design
Food contact equipment shall be constructed to be easily cleaned, using durable materials, resistant to multiple washings, and have a self-draining system when used in wet process areas. All equipment must be designed to minimize contact between the hands of operators and products. Piping systems should be constructed and maintained in order to be easily and periodically cleaned and have no dead ends.

## Product contact surfaces
Equipment and utensils in contact with food must: be made of washable, waterproof, corrosion-free, and nontoxic materials; be maintained in good condition and be correctly stored to prevent degradation; and have evidence of their suitability for contact with food.

## Temperature control and monitoring equipment
When equipment is used for thermal processes it has to be able to meet and maintain the established temperature during the process and to monitor and control the temperature (preferably automatically and continuously).

## Cleaning plant, utensils, and equipment
A cleaning program should be documented to ensure that all areas, utensils, and equipment are cleaned correctly and at a defined frequency. In Prerequisite 8 these documents are explained in more detail. When a cleaning–in-place (CIP) system is used (Prerequisite 8), it is particularly important to define responsibilities for the use and management of the system to guarantee its correct functioning. This system can be considered a sensitive area and managed according to Prerequisite 15.

## Preventive and corrective maintenance
The facility shall establish a preventive maintenance program that includes, at the least, all devices used to monitor and/or control hazards.[7] This program can include: definition of activities (e.g., mechanical or electrical verification); schedule of the maintenance (and calibration activities); responsibility definition; and calibration and verification activities (planning and procedures).

Preventive or corrective maintenance can be carried out by internal personnel or external specialized technicians. Both must have had previous training on the hazards that maintenance activities can represent for food products.

---

[7] The organization should consider extending the program to all equipment. The use of a record to list all equipment could be a good solution to better control maintenance and its schedule.

During maintenance, all risks of contaminating the product or adjacent equipment must be avoided. A person with adequate training and knowledge must be named responsible for verifying whether the equipment is properly sanitized before re-use.

All products used in maintenance or repair of equipment and that may be in contact with food products must be adequate (i.e., food grade).

## Prerequisite 6: Management of purchased materials
### General requirements

Before the acquisition of materials, an assessment of the suppliers shall be performed in order to ensure that they have the capability to meet specified requirements (i.e., whether the products they commercialize are safe and suitable for the intended use).

### Selection and management of suppliers

The standard defines the need to establish a process of selection, approval and monitoring of suppliers. This process should assess the risks and the suppliers' ability to meet food safety expectations, as well as the fulfillment of all requirements and specifications (defined by the organization or imposed by law). It should include a description of how suppliers are assessed and their performance monitored to ensure continued approval status. The organization should maintain an updated record of approved suppliers that includes all the contact information (e.g., address, telephone number, e-mail, name of the person to contact in case of emergency) necessary in cases of withdrawal/recall (Prerequisite 12), and data related to the supplier's ability to fulfill requirements (e.g., information requested, audit results, analytic or organoleptic results, customer complaints).

Selected suppliers should preferably have:
- food safety and/or quality certification (e.g., ISO 9001, ISO 22000, FSSC 22000, BRC, IFS, SQF);
- approval for supplying countries or groups of countries (e.g., the EU, Brazil, USA);
- good results on audits performed or requested by the organization; and
- good references.[8]

### Incoming material requirements (raw/ingredients/packaging)

Products must be checked before and during unloading operations to verify that their safety was not compromised during transportation. The tests to be carried out at the time of reception shall be defined and take into account the type of material (and the associated risk), the quantity supplied, the supplier's history and frequency of delivery, and whether the product has already been tested and verified by an external organization. Examples of tests that can be performed during reception

---

[8] Gather market information about the supplier: identify the competitors or customers which it supplies and ascertain their level of satisfaction and fulfillment towards that supplier (in particular regarding issues related to food safety).

include: temperature control at the moment of reception and/or during transit; label verification; expiration date and lot control; assessment of hygienic conditions of the vehicle; presence of foreign bodies; sensorial analysis (when applicable); and compliance with other specific requirements established with the supplier.

When the materials are not in conformity with the food safety requirements, they should be treated as potentially unsafe as described in Section 4.5.10.

The control of materials transported in bulk containers is more challenging and requires special attention at the moment of reception. Bulk materials can only be unloaded after control, verification, and approval. When the discharge is performed in piping systems, their access must be covered, closed, and identified.

### Prerequisite 7: Measures for prevention of cross-contamination
### General requirements
ISO/TS 22002-1:2009 states that a program capable of preventing, controlling, and detecting at least physical, microbiological, and allergen contaminations should be established.

#### Microbiological cross-contamination
Some areas are especially susceptible to cross-contamination either because they are too close to each other (e.g., no physical barrier between raw and finished products), or because they are located where pathways cross-over (e.g., waste removal and end-product circuits). A hazard assessment shall be carried out to identify potential sources of contamination, the impact they can have on products, and the necessary measures to control them.

Examples of these measures are: build physical barriers; define specific requirements to access certain areas (e.g., definition of specific clothing to be used when handling products ready to eat); set different times frames to avoid cross-over when it is not possible to ensure the separation of movement circuits of raw materials, finished products, waste, and personnel; and create air pressure differentials.

#### Allergen management
Allergens present in the product shall be declared on the label of finished products.[9] In the case of products intended for further processing, this information must be present on the label or accompanying documentation. The main focus of the organization should be to avoid the occurrence of cross-contamination.

Allergen cross-contamination may have two sources: conveyors or containers used by products with allergens, which are then used for other products without being properly sanitized; or through contact with ingredients or products in separated production lines. The use of water is not enough to avoid cross-contamination, and

---

[9] The organization must identify the statutory requirements applicable. Substances that can cause allergies or intolerances can be found in European Regulation (EU) No 1169/2011, in Public Law 108-282-AUG.2, 2004 (US) and in the *Codex Alimentarius* Commission (REV 1-1985 (1991)), for example.

the use of appropriate detergents is recommended for cleaning. One way to avoid accidental transfer of allergens is promoting the systematic cleaning of work surfaces and avoiding contact with other foods. Other techniques may include the handling of products in separate places, the use of specific materials for specific products (e.g., define color patterns to distinguish each material according to its application), a production planning that takes into account those risks (e.g., guarantee that allergenic products are the last product manufactured before cleaning), and the implementation of tests to verify the effectiveness of the elimination of allergens. Staff who handle allergenic material must be trained in the manufacturing practices.

### Physical contamination

Possible sources of physical contamination shall be analyzed and procedures to control them must be developed. All equipment used must be resistant and an internal audit to check its state should be carried out regularly. In order to avoid physical contamination, brittle materials, such as glass and plastic, should be avoided.

ISO/TS 22002-1:2009 identifies examples of sources of potential contaminations: wooden pallets, tools, personal protective clothing, and equipment. As examples of measures to prevent, control, and detect physical contaminations, the technical specification suggests the use of covers to protect equipment and containers, magnets, filters or screens, and metal detectors or x-ray detectors.

### Prerequisite 8: Cleaning and sanitizing
### General requirements

Organizations dealing with food products must implement cleaning and sanitizing programs. Their control and supervision depend on the size of the operation and on the nature of the activities. Programs that ensure monitoring of adequacy and effectiveness of procedures for cleaning and sanitizing should be implemented.

### Cleaning and sanitizing agents and tools

Facilities, equipment, and tools should be maintained in good condition to facilitate cleaning and sanitation. Chemicals and cleaning agents shall be clearly identified, stored, and used in accordance with the manufacturer's instructions.

Tools used for cleaning and sanitation should always be stored when not in use to prevent potential contaminations of food products. Tools and cleaning equipment shall be made of strong materials and maintained in a condition which does not represent a potential source of extraneous matter. The use of tools that may project waste or dust is not advised.

### Cleaning and sanitizing program

Cleaning and sanitizing programs shall be established and validated by the organization to ensure that all parts of the establishment are cleaned and/or sanitized. These programs (according to ISO/TS 22002-1) shall specify:
• areas and equipment to be sanitized;
• responsibility for the tasks specified;

- sanitizing method and frequency;
- monitoring and verification arrangements; and
- post-clean and pre-start-up inspections.

Other information that may be incorporated in these programs includes: cleaning products to be used; instructions for solution preparation; cleaning agents' contact time; utensils; and items of equipment that must be removed.

As well as having knowledge of the cleaning methods, the personnel responsible for cleaning must have adequate training on the correct and safe use of cleaning agents. The organizations shall provide the workers with mandatory protective clothing and the products safety data sheets (available as close as possible from the point of use).

In certain situations it may be useful to elaborate a schedule with set times for the cleaning of different equipment in order to prevent waste accumulation and/or cross-contamination.

A cleaning and sanitizing process may involve seven steps:

1. *Pre-cleaning*: This stage involves removing food and/or waste of larger dimensions from the area to be sanitized and protecting the parts of equipment that are sensitive to water.
2. *Pre-rinse*: At this stage, water is used to remove any waste that still remains on equipment.
3. *Cleaning*: This stage aims to remove food residues, dirt, grease, or other waste using a cleaning solution.
4. *Rinse*: At this stage all equipment should be rinsed with water to remove all traces of food and detergent.
5. *Disinfection*: The sanitizing agent shall be applied for a specific time. This procedure aims to reduce/eliminate microorganisms that are present on surfaces.
6. *Post-rinse*: At this stage all equipment is rinsed with clean water. This process is repeated until complete elimination of residues of each sanitizing solution.
7. *Cleaning efficiency verification*: The cleaning efficiency shall be controlled as appropriate.

At the end of cleaning, all equipment must be air-dried or dried with disposable/dried towels, and products/cleaning utensils shall be stored appropriately.

## Cleaning-in-place (CIP) systems

Cleaning-in-place systems are used to clean contact surfaces without dismantling the equipment and are generally considered to be faster, easier, and cheaper than manual cleaning. Parameters such as type and concentration of detergents, contact time, temperature, and flow shall be defined and monitored. The personnel responsible for controlling and monitoring these parameters shall have appropriate knowledge and training. The system must be monitored, maintained, and validated to guarantee the effectiveness of cleaning. The effectiveness should be regularly verified (e.g., control of detergents residues, analysis of the rinse water or product).

## Monitoring sanitation effectiveness

A cleaning and sanitation program shall be monitored at a frequency specified by the organization to ensure their effectiveness.

Apart from qualitative evaluation, which can be performed just after the cleaning processes by a designated person, quantitative tests should also be made periodically. However, these tests do not always provide conclusive and quick results. There are several tests that can be performed, the most important being those that determine whether any organic matter is left on the surfaces and/or assess the microbiological contamination of cleaned surfaces. In the latter case, it is necessary to define which microorganisms shall be investigated and establish limits for compliance. One method of classifying the cleanliness of equipment, utensils, and facilities is the number of microorganisms per $cm^2$.

## Prerequisite 9: Pest control
### General requirements

Inspection and monitoring procedures shall be implemented to avoid an environment favorable to the attraction and development of pests. For that reason, facilities should be maintained in good hygiene conditions and incoming materials shall be examined.

## Pest control program

Pest control may be managed internally by a nominated person or this service may be provided by an outsourced organization.[10] That person/contractor will be responsible for monitoring, detecting, and eliminating pests according to a program established for this purpose.

A pest management program shall be documented and should identify:
* controlled pests;
* methods and periodicity of controls (the time of year or other factors that may affect the occurrence of pests should be taken into account);
* location of pest control in the plan of the facilities; and
* list of chemicals approved in each area of the establishment.

The following documents must also be obtained:
* copy of service contract (in case of an external service provider) or evidence of the competence of the designated employee and definition of responsibilities;
* updated technical sheets and safety data sheets of chemicals used; and
* pest control records.

## Preventing access

In order to prevent pest access, buildings shall be maintained in a condition of good repair. Holes (e.g., gaps around pipes), drains, and other potential pest access points shall be sealed. External doors, windows, and ventilation openings shall be

---

[10] In this case, it is necessary to define a person to deal with the expert from the organization contracted.

designed to minimize the potential entry of pests. Periodically, the effectiveness of established control measures shall be evaluated (e.g., checking the conditions of mosquito nets, bait stations, insect traps, etc.). It is advisable that, as well as these controls, a more thorough inspection of pest activity is performed annually both inside and outside the facilities.

### Harborage and infestations

Storage facilities and established procedures should minimize the availability of food and water to pests. Whenever infested materials are detected, they shall be removed and placed in areas which ensure that they do not contaminate other materials, products, or buildings, and identified as potentially unsafe (Section 4.5.10).

The outside area shall be maintained under conditions which minimize the attraction of pests, such as the absence of:

- holes and undergrowth near the facility;
- waste left on the ground (e.g., cloths, paper, plastic films); and
- materials that are not used (to prevent degradation due to climatic factors or from becoming harborage places for pests).

### Monitoring and detection

As previously mentioned in 'Pest control program' of this prerequisite, detectors and pest traps shall be identified in the plan of the facilities. Their location shall take into account the type of pest, the activity carried out at that location, and historical information. The type of detector and trap must have the appropriate characteristics for their intended purpose and location.

The results of monitoring activities shall be recorded in order to allow the identification of events and trends, enabling the assessment of the effectiveness of the periodicity and location of detectors or traps.

### Eradication

Every time there are suspicions or evidence of infestation, eradication measures shall be put in place. When the use of pesticides to eradicate pests is necessary, it should be restricted to trained personnel and must be controlled to avoid risks to the operator or food products. If the possibility of equipment contamination exists, such equipment should be carefully cleaned before restarting the production activity.

Records of pesticides used shall be maintained to show: date and place of use; application method; target pest; type of pesticide; and quantity and concentrations used.

### Prerequisite 10: Personnel hygiene and employee facilities
### General requirements

The hygiene and behavior requirements of the personnel belonging to the organization shall be documented and be proportional not only to the type of activity that the organization carries out and to its position in the food chain, but also to the degree of risk the personnel may pose to the products and production areas. Different degrees of requirements may be considered within the same organization, depending on the probability and severity of contamination (e.g., different clothing or cleaning/changing periods). These requirements should also be respected by visitors and contractors.

### Personnel hygiene facilities and toilets

Sanitary facilities and changing rooms should be available and maintained to ensure the degree of personal hygiene required by the organization. These facilities must be clearly identified and located between the access point to the interior of the building and the access point to the production area. The sanitary installations shall not communicate directly with the production or packaging and storage areas.

Establishments must comply with the following requirements.

- Provide sufficient hand washing and drying equipment. Such equipment should supply hot and cold water, soap and disinfectant, wipes for drying hands and paper containers. Lavatories must have taps which are not operated manually.
- Provide a sufficient number of toilets (kept organized and in good condition).
- Provide adequate changing facilities for employees (with lockers for employees' personal objects).
- Ensure that in the path between the dressing rooms and production area the risk of workwear contamination is diminished (e.g., minimizing the distance and using a dedicated access).

### Staff canteens and designated eating areas

The canteens and the areas designated for the consumption of food must be located so that they are not a potential source of contamination to the production area. Food storage and preparation should also be conducted under hygienic conditions. When employees bring their own meal, it should be stored and consumed only in designated areas (never in changing facilities).

### Workwear and protective clothing

Employees who work in production or are exposed to products and/or materials must wear clean and appropriate work clothes (e.g., light-colored caps and coats without outside pockets above the waist and no buttons, plastic gloves, oversleeves and protective footwear which is nonslip and non-absorbent). Clothing should be appropriate to ensure proper protection of hair, beard, moustache, and sweat so they do not contaminate the product. Gloves should be made of suitable materials for contact with food and kept clean and in good conditions (the technical specification discourages the use of latex gloves). The work clothes must be designed to adapt to the different work areas and must be replaced periodically.

Visitors or others who attend the production area can only do so if adequately equipped in accordance with the organization requirements (the use of visitor's kits or equivalent equipment brought by the visitor is recommended).

## Health status

Employees who for the first time initiate the activity of manipulating food products should undergo a medical examination[11] to assess their ability to handle foodstuffs. This assessment must be renewed periodically as established by the organization or statutory obligation. The organizations shall keep evidence of this aptitude together with the contracts/agreements which establish the examinations made and their periodicity. It is suggested that a file is created to record each worker's medical examination status and plan. This information can be filed together with the worker training record (Section 4.1).

Employees suffering from infectious diseases of the digestive system (e.g., dysentery, typhoid fever, viral hepatitis A, hepatitis E virus), from diseases that affect food safety (e.g., active pulmonary tuberculosis, suppurative dermatitis), or employees with skin lesions that cannot be protected should be transferred to other places or activities that do not affect food safety.

## Illness and injuries

When employees with supervising responsibilities over operators who handle food products detect (or are informed) that an operators suffers from a condition that can constitute a hazard for food safety (e.g., jaundice, diarrhea, vomiting, fever, infected skin lesions, diseases transmissible to food), it must be reported to management and the operator excluded from tasks related to food manipulation, primary packaging materials, or food contact surfaces. Temporary workers, contractors, and visitors should also be made aware that their health status must not pose a risk to food safety.

Any injury, wound, or burn exposed should be protected with appropriate materials (e.g., dressings, bandages), preferably brightly colored, containing a metal detectable strip and, when possible, protected by gloves or clothes.

## Personal cleanliness

Employees who work directly or indirectly with food must maintain a high level of personal hygiene so that contamination of food is minimized. In the production areas the employees must wash and, when necessary, disinfect their hands, at least:

- before beginning any food handling activity or restart after smoking, eating, or drinking;[12]
- before putting on gloves;
- immediately after using the lavatory or blowing the nose;
- after handling any potentially contaminated material;

---

[11] The technical specification mentions that medical examinations may not be carried out in some countries due to legal restrictions.

[12] A way to ensure proper sanitation of workers when entering production areas is the installation of sanitary access control systems. These may possess, among other utilities, automatic washing and hand disinfection as well as shoe soles cleaning. It is common that these devices possess a turnstile door that prevents passage before the operator has completed the entire sanitation process.

- after handling waste and/or residues; and
- after change of clothing.

Nails should be kept clean and short. Hair should always have a clean look and be completely covered (e.g., hairnets). It is forbidden to cough, exhale, or spit on the materials, product contact surfaces, and products.

### Personal behavior

A policy that describes the behavior required in the areas of processing, packaging, and storage should be documented. This document could lay down rules on the following subjects:

- locations designated for smoking, eating, and drinking;
- use of personal ornaments such as rings, necklaces, earrings, piercings, watches, bracelets, and pins (taking into account the religious, ethnic, medical, and cultural issues);
- locations designed for storage and use of personal items;[13]
- prohibition of the use of nail polish, false nails, false eyelashes, and writing equipment behind the ear;[14] and
- organization and cleaning of personal lockers.[15]

## Prerequisite 11: Rework
### General requirements

Rework is the reuse of products from previous productions. These products must be stored, handled, and used in a way that maintains their safety, organoleptic characteristics, traceability, and compliance with documented requirements.

### Storage, identification, and traceability

Stored rework shall be protected from exposure to microbiological, chemical, and extraneous matter until the time of use. These products must be clearly identified and labeled to allow traceability. Traceability records for rework shall be maintained and ensure information such as: product name; production date; shift; line of origin; shelf life; lot; quantity; and justification (for rework).

### Rework usage

When rework is incorporated into a product as an 'in-process' step, its type, conditions of use, and the acceptable amount shall be defined in order to guarantee that those operations do not affect the product food safety requirements or even

---

[13] Medicines should be forbidden in food handling areas.

[14] The writing materials used should be appropriate; there are solutions on the market for using pens within the production area with properties that enable them to be detected by metal detectors and x-rays. These pens are impact-resistant so that food safety is not compromised in the event of their loss or damage. They should also be made of materials that meet the established requirements for contact with food and nontoxic paint.

[15] Lockers should be regularly checked to discourage misuse (e.g., storage of dirty clothes, food, or work tools).

organoleptic characteristics. The process step, method of addition and any necessary pre-processing stages shall also be defined. The need to change label content (e.g., new ingredients or allergens) must also be assessed.

When rework activities require the extraction of a product from its recipient, control measures should be established to ensure that packaging materials do not contaminate the product.

### Prerequisite 12: Product recall procedures
### General requirements
A procedure should be implemented to ensure the identification, location, and removal of products that do not comply with food safety requirements. To this end, the company must have a traceability system that allows it to track the movement of products through the different stages of production, processing and distribution (ISO 2007). More details are given on this subject in Section 4.5.10, including a proposal for the content of the procedure.

### Product recall requirements
A list of key contacts shall be maintained by the organization in the case of product withdrawal, including not only the customers but also suppliers and authorities. When products are withdrawn due to food safety hazards, the conditions in which they were produced shall be identified and the safety of other products (produced under the same conditions) should be examined. If food safety is at risk, those products should also be withdrawn. In any situation, the need to make a public warning should be considered.

The products withdrawn for being unsafe for human consumption should be clearly identified for disposal to prevent them from being placed on the market again.

### Prerequisite 13: Warehousing
### General requirements
The places and conditions in which materials and products are stored shall ensure that they remain clean, dry, well-ventilated, and protected from dust, fumes, odors, and other sources of contamination.

### Warehousing requirements
Temperature, humidity, and other environmental conditions should be controlled when required by product or storage specifications.

The storage of products should take into account the expiration date and ensure that the used/delivered product has the shorter expiration date. This way, organizations must manage the stocks according to the FIFO methodology (first-in-first-out) or FEFO (first-expire-first-out).

When the products are stacked, the need for measures to protect the lower layers shall be considered. Products should always be placed on pallets that must be distanced from walls and other pallets to allow cleaning and pest inspection.

Waste materials, chemicals, and nonconforming products should be stored in a separate area of products and materials. The use of forklifts powered by gasoline or diesel in product storage areas should also be prevented.

### Vehicles, conveyances, and containers

Vehicles, conveyances and containers shall be maintained in a good state of repair and be safe, innocuous, and cleaned to reduce the risk of food contamination. Prior to loading, it should be guaranteed that these conditions are met.

The organization shall implement procedures and install equipment that ensures temperature and humidity control where required by the characteristics of the products, customers, or regulatory authorities.

During transport, products should be protected from adverse conditions (e.g., direct sunlight, rain, temperature, humidity) and from impacts that may cause damage to the packaging and food. Bulk containers originally intended for use in the transport of food products cannot be used to transport nonfood products. If a vehicle was used to transport nonfood products, it must be cleaned before it is used to transport packaged foods.

### Prerequisite 14: Product information and consumer awareness

The organization should provide information relative to the food it produces to ensure a high level of protection of health and consumer interests, allowing consumers to make informed choices and use the food in a safe way.

There are several methods of transmitting this information, including websites, advertising, and labeling. However, the latter method is undoubtedly the leading vehicle to reach the consumer. In addition to the information required (by law or customer agreement), labels may also include storage, preparation, and serving instructions. The use of QR codes as described in Box 4.2 allows organizations to take advantage of so-called 'extended packaging.'

Companies can also promote or participate in actions to increase consumer awareness about the importance of storing, handling, and preparing food properly to ensure its safety. In 2001 WHO introduced the Five Keys to Safer Food poster and later (2006) a manual to spread throughout the world the message that: (1) food should be kept clean; (2) raw and cooked foods should be stored separately; (3) food must be cooked thoroughly; (4) food must be stored at a safe temperature; and (5) food handlers must use safe water and raw materials to guarantee food safety (WHO 2006).

### Prerequisite 15: Food defense, biovigilance, and bioterrorism
### General requirements

Each organization shall assess the potential danger of acts of sabotage, vandalism, or terrorism to their products. PAS 96:2014 *Guide to protecting and defending food and drink from deliberate attack* (Anon 2014) proposes a Threat Assessment Critical Control Point (TACCP) approach and the creation of a team to act in these cases. The TACCP team should re-assess every 2 years the probability of a threat occurring. The HACCP team of the organization shall inform the TACCP team whenever it considers that an abnormal result of laboratory tests on products or services may be caused by acts of sabotage.

### Access controls

The organization shall identify, preferably in the facilities plan, the areas considered more sensitive or susceptible to vandalism, sabotage, and terrorism. Access to these places should be denied to unauthorized personnel by using locks or electronic keys.

Employees should always be identified through ID card or working uniforms. The visits should be scheduled and justified, unless visitors belong to a recognized authority. During the visit, visitors must always be accompanied by an employee.

## 4.5.3 Preliminary steps to enable hazard analysis (Clause 7.3)

### General

Prior to making a hazard analysis, some relevant aspects must be taken into account for this process to be carried out comprehensively and properly. These aspects are described in the following sections. All documents regarding the hazard analysis must be collected, maintained, updated, and documented.

### Food Safety Team

The Food Safety Team should consist of senior management members and personnel from the organization's key departments (e.g., quality, logistics, production, trading, maintenance, human resources) in order to gather skills not only for food safety but also related to the processes, products, equipment, and human resources management. The composition of the team could be adjusted as and when required, making use of elements from other areas or external organizations. Documentation evidencing the skills of team members and supporting their presence in the team should be kept at all times.

### Product characteristics

### Raw materials, ingredients, and materials for contact with the product

This step allows an exhaustive review of the characteristics of such products with the aim of identifying potential hazards to the end-products or manufacturing processes and/or identifying preventative procedures. To conduct a hazard analysis, the standard identifies the following relevant points: biological, chemical, and physical characteristics; composition of formulated ingredients such as additives and processing aids; origin; production method; methods of packaging and distribution; storage conditions; expiration date; preparation and/or handling before use or processing; acceptance criteria related to food safety; and specifications of purchased materials and ingredients appropriate to their intended use.

**Box 4.6**   Defining risk associated with seafood products

The specific nature of seafood products, particularly of those that are obtained directly from nature, poses a great challenge in the knowledge and definition of risks associated with these products. As well as the intrinsic factors of the product and environment (e.g., origin, physical, chemical and microbiological characteristics of the aquatic environment, migrations), the safety of the fish product is also conditioned by the handling and transportation from the extraction or growth areas to the manufacturing companies.

In order to perform a successful hazard analysis, the organizations should seek to obtain from their suppliers all the possible information about the mentioned factors and proceed to a rigorous control at reception, not only from the point of view of physical and organoleptic characteristics, but also documental.

To support this analysis a similar approach to that carried out in FAO's Technical Paper 574 can be used, which categorizes the degree of risk of fresh seafood products and processed seafood products. In the particular case of fresh seafood products, fish was divided into six major groups (no terminal heat application; bad safety record; no CCP for identified hazard; harmful re-contamination; abusive handling; growth or accumulation of hazard) that were evaluated against five characteristics and risk factors. Upon the results of the assessment of each of these factors, fish is identified with high, medium or low risk. High-risk examples include live raw molluscan shellfish and tropical reef raw fresh/frozen fish and crustaceans. Low-risk examples include raw fresh/frozen fish and crustaceans other than tropical reef and scombroid fish.

For the organization to comply with this requirement it must keep in its possession the product data sheets containing the information requested and periodically check their update. Alternatively, it can prepare specifications which must define food safety requirements to be fulfilled and send them to suppliers so that they can put in writing a commitment to respect the defined specifications.

It is also necessary to identify the updated statutory and regulatory requirements related to food safety and to verify that they are fulfilled by suppliers (data sheets) or to prepare and update internal specifications. Box 4.6 provides further insights into the challenge of defining risks associated with seafood products.

### Characteristics of finished products

The characteristics of the products should be specified in documents. These documents may be grouped into categories according to their ingredients, processes, or hazards and may form the basis for the preparation of technical product sheets (ISO 2014). Table 4.6 lists the information that the standard considers mandatory to describe the characteristics of products, and some guidelines for the content of each characteristic.

As mentioned in Table 4.6, shelf life is part of the information required to characterize the end-products. Box 4.7 identifies the main factors to consider by organizations when defining shelf life.

**Table 4.6** Information needed to describe end-products

| Information | Description content |
|---|---|
| Product name | Description that identifies the product or group of products (use photograph when appropriate). |
| Composition | Identification of product constituents[a] (e.g., raw materials, allergenic substances, food additives). |
| Biological, chemical and physical characteristics | Description of product or group of products' specific characteristics that may have relevance for food safety. In addition to specifying the physical, chemical, and microbiological parameters (and their limits whenever applicable), the identification of other physical characteristics such as type packaging or physical state may be relevant. |
| Shelf life | Identification of the shelf life assigned according to the defined storage and storage conditions. |
| Packaging | Description of the packaging materials and their composition. In the case of printed packaging the type of printing and their constituents must be mentioned. |
| Labeling | Instructions for handling, preparation, and safe use of product (e.g., temperatures and cooking times, storage temperatures, thawing procedures).[b] |
| Distribution methods | Description of the methods used for the distribution of finished products and related food safety requirements. |

[a] It is advisable that the presentation of the composition of the products in this document complies with specific requirements for labeling;
[b] It is also possible to include other information in the label of the product that, although not related to food safety, is a mandatory statutory/regulatory requirement.

## Intended use

For the hazard analysis to be effective it is essential to make an assessment of the intended use of the manufactured products, whether in the subsequent stages of the food chain or with the final consumer. This review should document not only the intended/expected use but also any improper manipulations of products. A common example in this stage is the consideration of the improper use of products in not complying with the defined preservation temperatures or for insufficient thermal treatment. Another example is the use of the product for consumers intolerant to certain ingredients or with vulnerable immunological system (e.g., infants, elderly). Special attention should also be given to the identification of consumer groups to which the products are intended, particularly if these are especially vulnerable to specific food safety hazards. The information obtained may be displayed alongside the definition of the characteristics of finished products.

The result of these assessments may lead to: the consideration of new hazards; the implementation of new control measures; the changing of processes or product formulas; or communicating instructions for product use/preparation to the customer/consumer.

**Box 4.7**    Shelf life determination

Shelf life can be defined as the period during which the product maintains its microbiological safety and sensory qualities at a specific storage temperature. It is based on identified hazards for the product, heat or other preservation treatments, packaging methods, and other hurdles or inhibiting factors that may be used (CAC 1999). This definition describes the two key factors that organizations responsible for defining the shelf life should consider:
- the time until which the product loses organoleptic characteristics and no longer meets the expectations of customers/consumers; and
- the time until which the product becomes unsafe.

The assigned shelf life should be the lowest period of the two. However, this time is very conditioned not only by the characteristics of products, processes, and packaging (usually controlled or known by the organization), but also the circumstances under which the product is transported, stored (in customers or retail), and prepared for consumption (often out of the organization control). In this sense, beyond the communication of the shelf life organizations must define the conditions under which said shelf life is applicable, through information included in the label (especially when destined to the final consumer) or instructions for use and conservation defined in specifications and data sheets. However, it is advisable that the studies for the definition of shelf life also consider scenarios where products are subject to non-ideal conditions. These studies are time-consuming, complex, and costly and, particularly in organizations with fewer resources like small and medium-sized enterprises (SME), it may be advisable to resort to published literature, external experts, or legal authorities for support.

The two most common ways of indicating the shelf life are the 'use by' and the 'best before' dates. When you define the shelf life as 'use by,' it is indicated that after that date the product should not be consumed because it may constitute a danger to human health. In this case, as shelf life commonly corresponds to a short period of time, the loss of organoleptic characteristics is not so relevant. The 'best before' is a minimum durability date set by the manufacturer up to which, if the product is handled and stored in accordance with their instructions, the organoleptic and safety features are guaranteed.

## Flow diagrams, process steps, and control measures
### Flow diagrams

The *Codex Alimentarius* defines flow diagrams as a systematic representation of the sequence of steps or operations used in the production or manufacture of a particular food item (CAC 1969). Flow diagrams shall be prepared for all products covered by the FSMS. The use of flow diagrams allows:
- the visualization of the entire manufacturing process on a graphical form;
- an easier interpretation of particular production processes where these are complex; and
- an actual display of the positioning and sequence of all steps of the production process.

ISO 22000:2005 states how flow diagrams should be grouped by product categories or existing processes in the organization, and must contain:

- the sequence and interaction of all steps of the process;
- any external processes and subcontracted work;
- the steps of reception of raw materials, ingredients, and materials that come into contact with the product as well as the entry locations in the production flow;
- the completion of reprocessing and recycling and entry of intermediate products; and
- the release or removal of the finished products, intermediate products, by-products and waste.

In addition to the above points, flow diagrams may also include:

- existing control measures at every stage;
- identification of the steps defined as CCPs and OPRPs;
- time and temperature conditions whenever there is an intermediate storage step; and
- date and signature of the person responsible for the verification of flow diagrams on site.

The type of information and level of detail of a flow diagram may depend on the type of product, the complexity of the process, or even on the intended use. Organizations may choose to create flow diagrams for a particular step that includes all products, as may be the case for the reception of raw materials or distribution of the final product. Very specific flow diagrams may also be created due to the complexity of the process or product characteristics. It is common to use combinations of these alternatives, where the first and last stages are more generic flow diagrams and the middle stages comprise several (more detailed) flow diagrams, representing the entire operation when combined. It may be necessary and even practical to have flow diagrams with a detail level below what is required to carry out the hazard analysis, to present in nontechnical meetings of the Food Safety Team or even to customers who request them (Wallace *et al.* 2010). Figures 4.13–4.15 provide examples of flow diagrams. These flow diagrams only have illustrative purposes; they are not intended to represent any existing process and should not be considered sufficiently complete and comprehensive to be used in any organization.

### Description of process steps and control measures

In order to perform hazards analysis it not only is necessary to know the sequence of the steps required to manufacture the product, but also its function and objective. The standard indicates the need to describe existing control measures and any process parameters relevant to food safety, including those that result from regulatory and customer requirements. In order to facilitate hazard analysis, it is recommended that this information is presented together with the flow diagram (see Figs 4.13–4.15). This allows hazard analysis to be carried out in a more integrated manner, considering the interconnection and interdependence between different control measures.

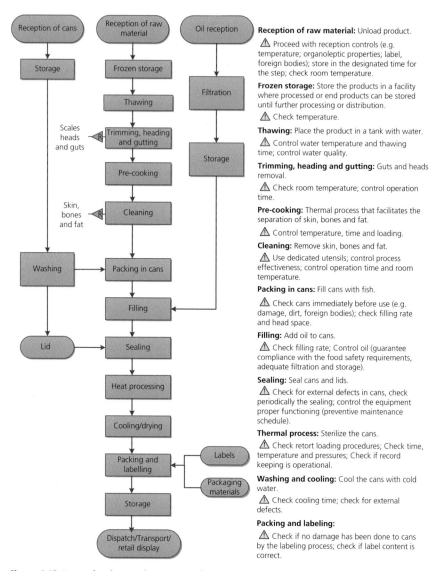

**Figure 4.13** Example of canned tuna manufacturing flow diagram.

## 4.5.4 Hazard analysis (Clause 7.4)
### General

The *Codex Alimentarius* Commission defines a hazard as 'a biological, chemical or physical agent in food, or the conditions in which they are with the potential to cause an adverse effect on health' (CAC 1969).

Hazard analysis is a process of collecting and evaluating information about the hazards and the circumstances resulting from their presence in order to decide

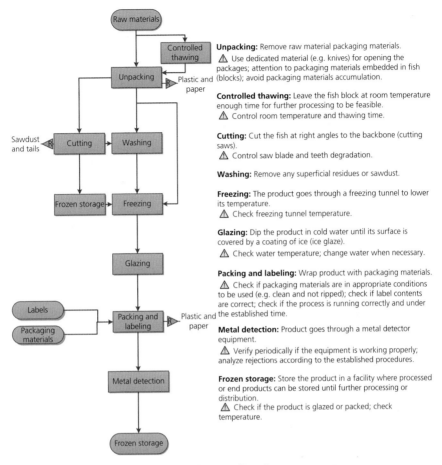

Figure 4.14 Example of frozen fish manufacturing flow diagram.

which are significant for food safety. This analysis can be performed for each category of products/processes and should be conducted by the Food Safety Team. All hazards should be considered, namely biological, chemical, and physical.

## Hazard identification and determination of acceptable levels

The Food Safety Team shall conduct a thorough search of all reasonably expected hazards affecting the safety of products. This analysis should cover all ingredients, raw materials, and materials that come into contact with the food product and process stages (preliminary stages, processing, and distribution).

According to the standard, hazard identification should take into account:

- the preliminary information and data collected in Section 4.5.3 ('Product characteristics');

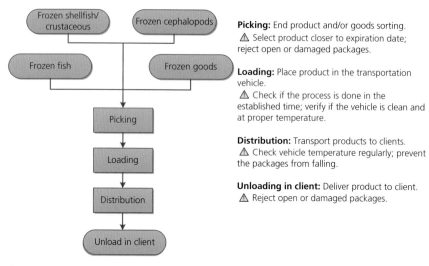

**Figure 4.15** Example of distribution flow diagram.

- experience (e.g., elements of the organization, sector specialists, or statutory and regulatory authorities); and
- external information (e.g., epidemiological data, historical data of the organization or sector, relevant guides or literature, and information from the food chain).

After identifying the hazards, the organization must determine their levels of acceptance. According to ISO 22004:2014 the acceptance level is set to be the permissible level of a hazard in the final product, which cannot be exceeded in order to ensure its safety, and its determination should consider (ISO 2014): established statutory and regulatory requirements; specifications established internally or by customers; and specific hazard information obtained internally or externally (especially considering the intended use of the product by the customer/consumer).

Once the levels of acceptance have been determined, the corresponding result and its justification must be registered.

### Hazard assessment

The hazard assessment step is intended to evaluate the hazards identified in the previous step that require control measures to achieve the level of acceptance defined.[16]

The standard states that each hazard should be evaluated according to its probability of occurrence and the severity of its effects. The probability should be determined taking into account the experience of the Food Safety Team in the

---

[16] The implementation guide for ISO 22004:2014 states that when a specific hazard is always maintained below the level of acceptance without any intervention from the organization, it must be defined as nonsignificant and does not need to be controlled.

**Box 4.8**   Probability of occurrence

> The definition of probability of occurrence of a risk and its fitting within levels defined by the food safety team is usually a big challenge during the hazard analysis. In fact, the objective definition of what is very likely or unlikely to occur and the justification for this decision may be controversial, not only within the food safety team but also towards external entities (e.g., customers, certification bodies, regulatory/statutory organizations). The use of historical occurrences within the organization, such as the number of annual incidences of a particular hazard, creates the possibility of making this analysis more objective. At the same time, it is possible to appeal to specialized publications such as reports produced by regulatory authorities or other organizations with information on events regarding such hazards. Appendix 2 contains an example of information that can be obtained from the Rapid Alert System for Food and Feed (RASFF) from the European Commission.

industry, records from previous incidents in the organization, and data obtained from the analysis of reports and studies published by relevant entities (Box 4.8). Severity is an intrinsic characteristic of the hazard related to the effect such hazard may have on consumer health, which reinforces the importance of having well-defined consumer groups (Section 4.5.3, 'Intended use').

The Food Safety Team should gather the necessary information for assessing hazards using scientific literature, databases, regulatory agencies, and industry experts. In this assessment it should also take the following into account (ISO 2014):

- the source of the hazard (its location and how it can be introduced into the product);
- the nature of the hazard (ability to multiply and produce toxins); and
- actions taken in subsequent stages of the food chain and their impact on the hazard acceptance level (organization end product).

According to the FAO, severity is defined as a degree of consequence which may result from the presence of a hazard. This organization classifies the severity in three levels (FAO 1998):

- **High:** serious health effects, life-threatening to the consumer. Examples of bacteria which may cause these effects include *Clostridium botulinum*, *Salmonella typhi*, *Listeria monocytogenes*, *Escherichia coli* 0157:H7, *Vibrio cholerae*, *Vibrio vulnificus*, paralytic shellfish poisoning, and amnesic shellfish poisoning.
- **Moderate:** Severe or chronic hazards to consumers, which may be caused by *Brucella* spp., *Campylobacter* spp., *Salmonella* spp., *Shigella* spp., *Streptococcus* type A, *Yersinia enterocolitica*, hepatitis A virus, mycotoxins, or ciguatera toxin, for example.
- **Low:** minor or moderate hazards to the consumer, such as symptoms caused *Bacillus* spp., *Clostridium perfringens*, *Staphylococcus aureus*, Norwalk virus, most parasites, and histamine-like substances.

It is common for organizations to use a risk matrix in the hazard assessment (Table 4.7). These matrices can range from simple to more complex, depending on the number of established probability and severity levels. However, the definition of many levels does not necessarily make the assessment of the degree of risk more accurate or correct. The most important is the substantiation used by the Food Safety Team to support their decisions and the definition and maintenance of a consistent methodology throughout the analysis. Box 4.9 contains an interview with Dr Thomas Ross, where the subject of risk assessment in the seafood industry is discussed.

**Table 4.7** Example of matrix used in hazard assessment

| | | Low (1) | Moderate (2) | High (3) |
|---|---|---|---|---|
| **Probability** | High (3) | 0 | 3 | 6 | 9 |
| | Moderate (2) | 0 | 2 | 4 | 6 |
| | Low (1) | 0 | 1 | 2 | 3 |
| | Neglegible(0) | 0 | 0 | 0 | 0 |
| | | Low (1) | Moderate (2) | High (3) |
| | | | **Severity** | |

**Box 4.9**   Risk assessment in the seafood industry

Dr Thomas Ross is a microbiologist who has developed expertise in 'predictive microbiology' and its application to quantitative microbial food safety risk assessment. He has been involved in numerous FAO/WHO expert panels to develop guidelines for microbiological food safety risk assessment and to use it to answer food safety management questions being considered by the *Codex Alimentarius* Commission, as well preparing microbial food risk assessments for both government and businesses in Australia. He has contributed to risk assessments concerning risk management of histamine in fish, histamine in Asian fish sauces, and risk ranking of hazards in seafood, which led to the development and publication of a simple risk assessment model (now known as 'Risk Ranger') that has also found utility in a number of other food safety risk assessments.

**What is the background to risk assessment in the seafood industry?**
The ideas of 'risk assessment' have been formally applied to microbial food safety since the late 1980s with impetus from the World Trade Organization (WTO) which mandated that risk to consumer health is one of the only legitimate reasons for restriction of free trade between nations, including seafood. In fact, one of the first international food safety risk assessments concerned seafood in international trade (FAO 1999) and others have been used to change international regulations concerning histamine levels in Asian fish sauces (CAC 2011).

**What are the main challenges and limitations to seafood risk assessment?**
Among the challenges to the widespread implementation of 'risk assessment' is the perception that it is labour- and data-intensive, and mathematically complex. But this is not necessarily correct: while the fundamental principles of food safety risk assessment have been articulated (CAC 1999), there is no single method prescribed to assess microbial foodborne risk. Rather, food safety risk assessment is simply a logical, structured and transparent approach to making a decision by systematically combining all the relevant information.

Risk assessment aims to combine data and knowledge about factors that influence a risk (e.g., foodborne illness) to: (1) help to evaluate the magnitude of the risk; (2) identify the factors that most affect the risk; and, potentially (3) identify the most effective way(s) to manage the risk to an acceptable level.

Nonetheless, a limitation to food safety risk assessment currently is the type and quantity of data needed to evaluate food safety risks, particularly for so-called 'boat-to-throat' risk assessments that consider how risks arise and change across all stages of the seafood supply chain, including variation in: (1) pathogen levels in products; (2) pathogen growth rates and limits; (3) handling, processing, and food preparation that affect pathogen inactivation and growth; and (4) amounts eaten by different groups of consumers and the susceptibility of those various consumers to harm from the pathogen.

Another issue in employing formal risk assessment is whether the cost of a fully quantitative risk assessment outweighs the benefits that might be achieved by that approach. For example, imagine a full risk assessment that requires 20 person-years of work to achieve the 'best' answer, and that the risk management strategy would cost US$ 2 million each year in testing. Imagine also that there is another risk management option based on a simpler risk assessment that achieves the same level of consumer protection and costs producers only US$ 200,000 per year in testing but a further US$ 200,000 per year through incorrect rejection of acceptable product. The less precise but more wasteful and 'conservative' strategy, while costing more in lost product, is actually more cost-effective overall, particularly for the initial cost of the risk assessment. Risk assessment methods used should therefore be consistent with the purpose and importance of the decision that needs to be made.

### How can HACCP benefit from risk assessment?

Development of HACCP plans also involves assessment of hazards (Principle 1), that is, 'the things that could go wrong.' The ideas of risk assessment are very apposite in this aspect of HACCP planning because, potentially, there are many hazards. Not all, however, are equally likely to occur and the consequences (e.g., number of people who become ill and how ill they become) of the presence of the hazard can vary enormously. Given the limited resources usually available for food safety management, it makes sense to use those resources to minimize the greatest risks first, that is, those that most need to be controlled through implementation of CCPs. Eminent food microbiologists have considered this in greater detail (Notermans & Mead 1996; Buchanan & Whiting 1998) and in the USFDA's Food Safety Modernization Act, introduced in 2011, the concept of risk-based CCPs was formally introduced in USA and termed hazard analysis risk-based preventive controls (HARPC). The idea of HARPC is to assign relative importance (as assessed by risk) to potential hazards and then to assign resources and strategies accordingly to control the most important hazards.

### Can you explain better the principles behind 'Risk Ranger'?

Food safety risk assessment seems to be an idea that will increasingly become a normal part of food safety management, including in the seafood industry. However, regulatory authorities are aware that risk assessment, while a simple concept, is not easily performed, particularly for microbial food safety hazards. That is because the food safety risk can change dramatically over time as microbial hazards grow or, if they are killed, as a result of the storage and handling of the food. For this reason, various organizations are attempting to provide resources to make food safety risk assessment more accessible to a wider range of users. The United Nations' Food and Agriculture Organization (FAO) and World Health Organization (WHO) have, on behalf of the *Codex Alimentarius* Commission, developed guidelines for practical implementation of microbial food safety risk assessment (see the JEMRA website, http://www.fao.org/food/food-safety-quality/scientific-advice/jemra/en/). Similarly, FAO's Fisheries Department have developed a practical guide for application of risk assessment in seafood industries (FAO 2004) that advocates the use of a simple risk assessment tool that has become known as 'Risk Ranger' (Ross & Sumner 2002; Anon 2015).

Risk Ranger is a simple spreadsheet software that guides users through the questions needed to assess risk, and then automates the calculations required to estimate the risk. It is a very useful tool for teaching the ideas of risk assessment but, for it to provide meaningful risk estimates, requires that the user has extensive relevant knowledge of risk-affecting factors. Similarly, iRisk (USFDA/CFSAN 2015) provides a generic, web-based, risk assessment model that guides users through the risk assessment process. As for Risk Ranger, the data required includes the food and its associated consumption data and processing/preparation methods, the hazard and knowledge of levels normally required to cause human illness, and the anticipated health effects of the hazard when ingested by humans. Clearly, some of this information will not be readily available to nonspecialist users and there are initiatives to establish publicly available web-based databases with relevant information. An example is the FoodRisk.org website (www.foodrisk.org) developed and maintained by the US FDA and Food Safety and Inspection Service.

Because so much data are needed to quantify the absolute risk from a particular food/process combination, sometimes a comparative approach is used to quantify the *change* in risk, or relative risk, due to some change in the process or distribution pathway. This simpler approach may be all that is required to make the required risk management decision, but usually requires far less data and perhaps less detailed knowledge and new modeling. Often an existing model can be used to determine relative changes in risk that would be expected from some new intervention (e.g., a change to the critical limits of a CCP).

**How do you anticipate the future of risk assessment?**
Risk assessment is a way to use available data and knowledge to make better decisions about risk management needs and priorities. It can clearly be an important aid to HACCP plan design to optimize use of risk management resources, and new HACCP approaches that involve risk assessment (rather than hazard identification only) are being formalized. While some forms of risk assessment are very complex and labor-intensive the principles of Quantitative microbial risk assessment (QMRA) can equally be applied, using simpler approaches, to the benefit of industry. Organizations should realize that food safety risk assessment is not only useful for setting national and global food safety criteria and for international trade negotiations; it can equally be applied using simpler approaches to the benefit of industry. Resources to support a wide range of food, and seafood, risk assessment decisions are already available, and more will continue to be added in the public domain to assist industry develop safer processes and protocols.

## Selection and assessment of control measures

For all the control measures defined in Section 4.5.3 ('Flow diagrams, process steps, and control measures') and according to their effectiveness against food safety hazards, the Food Safety Team should select the necessary combination to prevent, eliminate, or reduce food safety hazards to acceptable levels.

The control measures specified should be classified regarding the need to be managed by OPRPs or by the HACCP plan. A systematic approach for this classification is suggested and a flow diagram, such as the one presented in Figure 4.16, can be used by organizations to identify whether the control measures should be managed by PRPs, OPRPs, or CCPs.

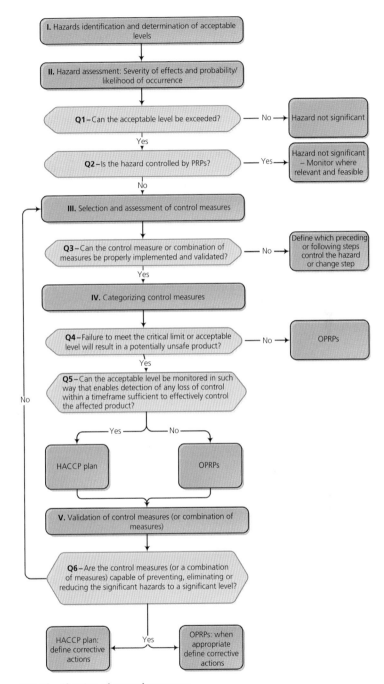

**Figure 4.16** Classification of control measures.

## 4.5.5 Establishing the operational prerequisite programs (PRPs) (Clause 7.5)

ISO 22000:2005 defines an operational PRP as a PRP identified as essential (by hazard analysis) to control the likelihood of introducing food safety hazards or the contamination or proliferation of food safety hazards in the product(s) processing environment (ISO 2005a).

The OPRPs manage control measures in a very similar manner to the HACCP plan. In some cases, OPRPs manage control measures that cannot be controlled by the HACCP plan; in ISO 22004:2014 it is mentioned that the control measures managed by HACCP plan should have critical limits and be monitored so that any loss of control is detected in sufficient time. The control measures identified which do not have these features cannot be managed by HACCP, but may be managed by OPRPs.

The technical specification ISO 22004:2014 introduces the concept of *action limit/action criterion* defined as measurable or observable criterion for monitoring a control measure managed by an operational PRP. An action limit/action criterion determines whether the control measure is under control or not, establishing the distinction between acceptable and unacceptable (depending on whether or not such a limit is met/achieved), thus indicating if the control measure is operating as intended. This concept is equivalent to the critical limit associated with CCPs. The difference is that, in the case of loss of control although the implementation of corrective actions is mandatory it does not result in potentially unsafe products. Operational PRPs shall be documented and include the following information (ISO 2005a).

- Food safety hazards to be controlled: describe the hazard and identify the step (set a numbering system to allow its identification in flowcharts, for example).
- Control measures: describe the measure or combination of control measures established to control the hazard.
- Monitoring procedures: establish the action limits or action criterion for the control measure, determine how often monitoring is performed, and define the person responsible for this task.
- Corrections and corrective actions: describe the actions to take whenever the action limits or action criterion are exceeded and define the person responsible for those actions.
- Monitoring records: identify the record of the food safety management system in which the control of the OPR is performed.

## 4.5.6 Establishing the HACCP plan (Clause 7.6)
### HACCP plan

According to the *Codex Alimentarius*, the HACCP plan is a document elaborated in accordance with the principles of HACCP to ensure the control of hazards that are significant for food safety (CAC 1969). Its establishment should be performed by the Food Safety Team and its effectiveness is essential to the fulfillment of the preventive nature of HACCP. In fact, if all CCPs are properly monitored and any deviation from a critical limit is quickly identified, the potentially unsafe product production is reduced and the need to control the final products is significantly restricted. The standard identifies the information that should be included in the document that manages the HACCP plan and, except for the identification of the critical limit, it is the same as described in the previous section for the operational PRPs.[17]

## Identification of critical control points (CCPs)

A critical control point is defined by the standard as a 'step at which a control measure can be applied and is essential to prevent or eliminate a food safety hazard or reduce it to an acceptable level' (ISO 2005a).

The definition of control measures results from the analysis conducted by the Food Safety Team of the steps of the process, which should subsequently be evaluated for the need to be managed by the OPRPs or HACCP plan (Section 4.5.4). This decision is often facilitated by the use of decision trees, as depicted in Figure 4.16. The ISO 22004:2014 establishes criteria to be met by the control measures associated with CCPs: the critical limit should establish the boundary between a safe product and a potentially unsafe product and its monitoring should allow timely actions to be taken to ensure acceptable levels of risk in the final product (ISO 2014).

Failure to comply with the critical limits will result in potentially unsafe products.

## Determination of critical limits for CCPs

Critical limits must be designed to ensure the control of hazards in food safety and should be specific and validated for each CCP. When a CCP controls more than one hazard, the most stringent limit from those determined for each hazard should be applied (ISO 2014).

These limits should be measurable, preferably by the monitoring of objective values such as temperature, time, humidity, water activity, etc. When limits are based on subjective data such as sensory, visual, and textural parameters, they must be supported by instructions or specifications accompanied by images to facilitate the perception (FAO & WHO 2009).

The basis for the definition of a critical limit must be documented in the food safety system. It is frequent for organizations to establish critical limits based on legislation and regulations of the sector but, in certain situations, it may be prudent to establish more stringent limits conditioned by the organization's positioning in the food chain and also by the knowledge of the common practice of manipulation or incorrect use of the products.

---

[17] This description also includes the concept *action limit/action criterion* (introduced by the 2014 revision of ISO/TS 22004 and not present in ISO 22000:2005), which should be replaced by the *critical limit* concept when applying the HACCP plan.

## System for the monitoring of CCPs

The implementation of a monitoring system is decisive for verifying that critical control limits do not suffer deviations. Whenever possible, this system should allow the monitoring of trends in control loss, thus allowing process adjustments to be performed before the critical limit is exceeded. The information derived from the monitoring systems should be evaluated by a designated person with the authority to take corrective actions when necessary. The frequency of monitoring should be sufficient to guarantee that the CCP is under control and to prevent the affected product from being used or consumed. Physical and chemical measurements (e.g. pH, temperature, relative humidity, time) are preferred since they allow immediate results, whereas microbiological parameters are often more time-consuming and therefore slower to obtain. All equipment essential to food safety must be calibrated. The responsibility and authority for the monitoring and recording of CCPs should be defined and the evidence of these activities should be filed (FAO & WHO 2009).

## Actions when monitoring results exceed critical limits

When there is a deviation from a critical limit it means that the product is potentially unsafe. All necessary measures should therefore be taken to ensure that the nonconformity is identified, the CCP put back under control and the reoccurrence of the deviation prevented. In this sense the HACCP team should specify (in the HACCP plan) the corrections and corrective actions that the organization shall take whenever critical limits are exceeded.

ISO 22000:2005 defines correction as an action to eliminate a detected nonconformity. An example of a correction is the reprocessing of a product under conditions that meet the critical limits or the replacement of incomplete or incorrect labels. A corrective action is defined as an action to eliminate the cause of a detected nonconformity or other undesirable situation. This is taken to avoid their repetition and implies a cause analysis. A corrective action is, by nature, more difficult to plan because it depends on the causes that lead to nonconformities. A common corrective action is to carry out training sessions to correct nonconformities resultant from a lack of information or employee training (ISO 2005a).

## 4.5.7 Updating of preliminary information and documents specifying the PRPs and the HACCP plan (Clause 7.7)

After establishing the OPRP and/or the HACCP plan it is necessary that the organization assesses the need for updated documentation and assumptions used in their elaboration, in particular (ISO 2005a): product features; intended use; flowcharts; process steps; and control measures.

In fact, this need can also arise from the common organization of activities such as changes in products/processes (e.g., creating new products or changing specifications thereof, purchasing equipment) or external changes to the organization (e.g., new statutory and regulatory requirements).

After updating the documentation, the organization shall assess the need to change the procedures and instructions established for the prerequisites, the OPRP, and/or HACCP plan.

## 4.5.8 Verification planning (Clause 7.8)

The concept of verification is defined by the ISO 22004:2014 as a confirmation by objective evidence that a specific requirement was satisfied. This confirmation can be accomplished through various activities conducted during or after operations in order to assess whether the control measures are implemented as planned, and should cover the whole food safety management system.

According to a list provided by the standard, verification activities should ensure that (ISO 2005a):

- the PRPs are implemented (e.g., periodic evaluation through audits or checklists);
- inputs for hazard analysis are continuously updated (e.g., assessment of changes in internal and external documents);
- operational PRPs and the elements contained in the HACCP plan are implemented and effective (e.g., verification of records, analysis of nonconformities, and measures implemented);
- hazard levels are within the defined acceptable levels (e.g., review of the results from monitoring acceptable levels); and
- other procedures required by the organization are implemented and effective (e.g., verification of hand-washing procedure through visual control and/or microbiological assessment of effectiveness; verification that equipment is calibrated as planned).

ISO 22004:2014 defines that the established verification plans should include the following information (ISO 2014):

- purpose (e.g., verification of the implementation of PRPs and their effectiveness);
- scope of the verification (e.g., control of documents and implementation);
- verification method (e.g., internal audit and on-site inspection);
- fequency[18] (e.g., twice a year);
- actions to be taken if the results are unsatisfactory (e.g., training on record filling, performing maintenance on equipment, segregating potentially unsafe products); and
- reporting requirements in case verification results are unacceptable (e.g., contact the team leader/Food Safety Team).

The application guide of ISO 22000:2005 places particular emphasis on the verification of the specifications of materials and contracted services, indicating

---

[18] The verification should allow the identification of failures in compliance with requirements set by the food safety management system. In this sense, its frequency should consider the history of verification activities, the impact on food safety of the procedure being verified, and the likelihood of the procedure modification.

that they must be defined and available for verification, especially when compliance with the risk levels of ingredients or raw materials is essential to ensure food safety (ISO 2014).

Verification activities shall be recorded and communicated to the Food Safety Team, which should carry out its analysis as described in Section 4.6.4.

### 4.5.9 Traceability system (Clause 7.9)

Traceability systems have been used for many years in various sectors. The growing consumer demand for information about the products they consume, as well as the increased size and complexity of the food chain, have created the need for organizations to implement traceability systems that allow the transmission of that information.

ISO 9000 presents a generic concept of traceability, defining it as the ability to trace the history, application, or location of the object in question. The *Codex Alimentarius* Commission, the ISO 22005, and the European Union (Regulation No. 178/2002) have very similar definitions in which they apply this concept to food products (Ryder *et al.* 2014). The standard ISO 22004:2014 considers traceability a basic tool for food safety, defining it as the capability of the organization to accompany their final products, raw materials, packaging materials, and ingredients throughout the food chain (ISO 2014). The fact that ISO has published the international standard ISO 22005:2007 *Traceability in the feed and food chain — General principles and basic requirements for system design and implementation* reveals the importance of the issue for the food industry and also the need to harmonize and guide the implementation of these systems (ISO 2007).

Box 4.10 introduces two examples of organizations that are enhancing confidence in their products by providing traceability information to customers.

The traceability system can be divided into internal and external traceability. The former tracks the transformations of the product inside the organization, while the latter allows the identification of suppliers and customers one step back and one step forward. This system is especially relevant when failures occur (e.g., if there is a food poisoning outbreak, the organization must have the information necessary to remove the product from the market and report to the competent authorities). When the cause of occurrence does not have its origin in the organization, the authorities use the information to try to identify the problem in the previous step of the food chain. A fully implemented traceability system involves the use of organization resources but, on the other hand, in the case of a withdrawal it can minimize the amount of product to recall and reduce the damage to the company's image.

A period for maintaining the records of traceability should be established taking into consideration the characteristics of the product, its expiration date, and the intended use by the customers/consumers (e.g., if it is to be incorporated in another product).

**Box 4.10**   Traceability

According to a study by Oceana (Oceana Study Reveals Seafood Fraud Nationwide) where, during 2010–2012, 1215 seafood samples (collected from 21 different states from the United States of America) were genetically identified, 33% of the products tested were considered mislabeled (Warner *et al.* 2013). Another threat to consumer confidence and even health is the illegal, unreported, and unregulated (IUU) fishing. This problem has grown in the last two decades and it is estimated to be of a scale of 11–26 million tons each year (FAO 2014). This fraudulent behavior represent a serious detriment to the consumer and promotes a feeling of distrust in the supply chain. Fish distribution systems are growing more complex and globalized, and it is becoming more costly to obtain information on the origin of and the processes that the fish has been submitted to until it reaches the consumer. The combination of these factors has encouraged the search for solutions to the development of systems that improve the traceability and legitimacy of information that is transmitted to the consumer.

The company Backtracker (Traceability and Verification of Seafood Products) is developing a database that organizations responsible for fish processing can use to record the fish route. This way the consumer can check the origin and stages along the food chain of the fish being consumed. Norpac Fisheries Export implemented a barcode system that allows customers to access a high level of detail about the product. Each fish shipment comes with a label printed with a QR code (see Box 4.2) which, when scanned, opens a link to a webpage of information, such as commander of the ship, when and where the fish was captured, and even the equipment utilized. This technology allows the promotion of fresh fish and offers the consumers a receipt that documents the origin of the fish they are buying. FishPrint® is other traceability program developed by Profish®, allowing information about harvest zone, vessel and fisherman, latin genus, and DNA test results to be accessed via QR codes.

## 4.5.10  Control of nonconformity (Clause 7.10)
### Corrections
As already mentioned in Section 4.5.6 ('Actions when monitoring results exceed critical limits'), a correction is an action to eliminate a detected nonconformity and should therefore be carried out immediately after its detection. The greater the elapsed time between the loss of control and the implementation of corrections, the greater the amount of potentially unsafe product. The affected product should be identified and evaluated and handled in accordance with the procedure for handling potentially unsafe products as described in Section 4.5.10 ('Handling of potentially unsafe products') (ISO 2014).

Corrections must be approved by competent personnel previously appointed to this function, and filed together with information on the nature of the nonconformity, the amount and lots of the affected products, and evidence of action taken (e.g., computer records, photographs, records of microbiological or physico-chemical analysis) whenever possible.

### Corrective actions
Corrective actions (also mentioned in Section 4.5.6, 'Actions when monitoring results exceed critical limits') are actions to eliminate the cause of a detected non-conformity or other undesirable situation in order to prevent its reoccurrence,

and must be initiated when critical limits or (when possible) an action/criterion limit are exceeded. Corrective actions are not only initiated as a result of corrections; any detected nonconformity should generate corrective actions even if a correction has not been made or considered necessary. Following the identification of the causes of nonconformities, and taking into consideration their nature, the possibility of withdrawal of the unsafe product should be considered.

A period to assess the effectiveness of the corrective action should be determined. They should only be considered effective if, during this period, there is no repetition of the nonconformity with origin in the same detected causes.

Standard ISO 22000:2005 requires that a documented procedure is established for the implementation of corrective actions. In addition to the actions described (Fig. 4.17), the documentation must indicate the person responsible for their implementation. It should be demonstrated that the nominated employees have the required skills to perform those actions (ISO 2005a).

The organization must have a record to document the entire process of managing the nonconformity, which can be divided into three main phases as follows.

1. *Problem identification*: Record the source of the problem (e.g., internal/external audit, OPR or CCP), the date of occurrence, and its description.
2. *Identifying the causes*: Register the causes that led to the nonconformity. To facilitate this determination, it is possible to use some methodologies already established for this purpose (Box 4.11 contains examples of such methodologies).
3. *Definition of measures and evaluation of their effectiveness*: Describe the measures identified as necessary to restore conformity (correction) and/or eliminate the causes (corrective action), those responsible for executing them, and the expected date for completion. Identify the period during which the effectiveness of corrective actions is assessed and whether the nonconformity has been corrected; otherwise, restart the process of identification of causes or take further corrective actions.

At all phases it should be clear who is responsible for implementation and/or approval, since nonconformity process can be managed by different persons.

## Handling of potentially unsafe products
### General
A nonconformity is understood as the failure to meet legal, regulatory, or internal procedures or requirements. This concept is related to the production of unsatisfactory results and potentially unsafe products (e.g., control loss in OPRPs and exceeded critical limits).

The organization shall establish appropriate procedures to prevent the introduction of potentially unsafe products in the food chain, unless (ISO 2005a):
- the organization is able to reduce to acceptable levels the risk to food safety or evidence that these will be reduced before entering the food chain; or
- despite the nonconformity, the product is still within the defined acceptable level.

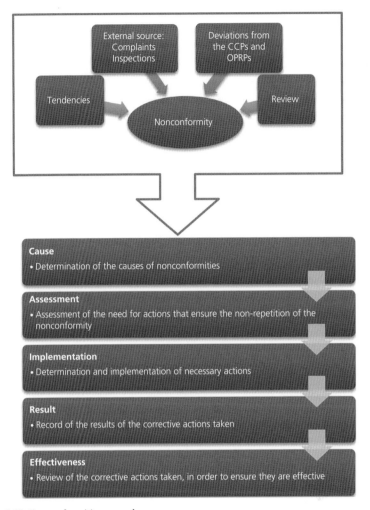

**Figure 4.17** Nonconformities procedure.

All products that may have been affected by the nonconformity must be clearly identified to prevent them from being used or marketed. If some of these confirmed unsafe products have left the organization, the withdrawal process should be initiated.

### Evaluation for release
The assessment for release is performed to ensure that a potentially unsafe product is only released to the food chain when it complies with the specified acceptance levels. For this purpose the evaluation must be conclusive and supported by evidence that ensures that the product is safe.

**Box 4.11**    Examples of methodologies to identify causes

1. *Cause–effect diagram*: This technique was originally proposed in 1976 in the book Guide to Quality Control (Ishikawa 1986). It is a technique that shows the relationship between effect and possible causes. It is schematically organized as a fishbone, allowing the visualization of the possible causes of a problem and analysis of the improvement processes. It uses the brainstorming methodology and classifies the causes in five groups (5M's): machine, material, method, man, and milieu.

2. *Brainstorming*: This is a technique for generating ideas oriented towards solving group problems, stimulating their creativity and the production of innovative ideas. It is divided in two stages. In the first, the goal is to unrestrictedly generate the biggest number of ideas possible to proceed to a second phase in which they are analyzed, discussed and selected. One of the basic rules of the first stage is the prohibition of debate and criticism of the ideas presented so that everyone has an equal opportunity. These rules mean that the parties have the freedom to say what they want without feeling inhibited by criticism.

3. *Tree of whys*: This is a technique which consists of placing a set of questions to the problems and seeking possible solutions. It is considered that the source of the problem is found when is no longer possible to find answers to the questions arisen (i.e., when there is no explanation for the event). An example application of this technique in the case of a nonconformity detected during transportation of products under controlled temperature is included below.

*Tree of whys example:* **Q: Why do products not reach customers at the right temperature?** A: Because the temperature during transport increases. **Q: Why does the temperature increase during transport?** A: Because during transport the doors remain open for too long. **Q: Why do the doors remain open for too long?** A: Because distributors load the vehicle incorrectly (disordered). **Q: Why do distributors perform a disordered load of the vehicle?** A: Because distributors lack training (and time) at the moment of loading.

According to ISO 22000:2005, the products can only be released when there is other evidence, in addition to the monitoring system, that proves that control measures or the combination of them were effective (e.g., able to ensure the desired levels of acceptance) or that the results of verification activities (e.g., laboratory analysis) testify that the affected product complies with the identified acceptable levels (ISO 2005a).

### Disposal of nonconforming products

When the assessment for release presents unsatisfactory results and the product is not acceptable for release, it shall be subjected to one of the following activities (ISO 2014):

- reprocessing or new processing of the product inside or outside the organization, in order to ensure that the food safety hazard is eliminated or reduced to acceptable levels;
- reformulation or usage for other purposes for which the product is acceptable (e.g., use in products which have an intended use that ensures their safety such

as mandatory thermal processing or changes in the composition of the product to ensure that it meets the requirements for food safety); or

- destruction.

## Withdrawals

The withdrawal concept refers to any measure aimed at preventing the distribution and exhibition of a hazardous product and its supply to consumers (EC & EP 2002). The withdrawal process is performed when a product is assessed as unsafe and the organization identifies that it is already in the next step of the food chain. As mentioned in Section 4.5.2, Prerequisite 12, the need to remove products produced under the same conditions should equally be evaluated.

The product withdrawal process may be generated when the product is not safe for human consumption (e.g., microbiological contamination) or does not fulfill the legal requirements (e.g., labeling).

When a nonconformity is detected and withdrawal conducted, as much information as possible should be gathered about the product including (at least): full product description including the identification of the lots of raw materials and packaging materials; product lot and quantity produced; date of manufacture and expiration date; and other information related to the product or process (e.g., identification of the production line and manufacturing time and manufacturing operators responsible for it).

For the recall of products to be complete and on time, the top management should nominate personnel to start and manage the withdrawal process. Traceability is a fundamental process for the withdrawal to be completed. This system is able to identify incoming material, as well as the initial distribution route of the finished product. The organization shall establish and maintain a documented procedure for the management of withdrawals which should include the following:

- the interested parties to be notified of the occurrence (e.g., statutory authorities, customers, consumers);
- actions to be taken at each stage;
- the treatment method of products/lots both withdrawn and still in stock and the process to keep them safe and under proper supervision until assessment;
- the method by which the withdrawals are documented (e.g., causes, size, results) and which ensures that they are included in the management review (Section 4.3.8); and
- procedures for verifying the effectiveness of the withdrawal program.[19]

---

[19] The organization must be able to demonstrate that all planned steps are met and allow for the withdrawal of any unsafe product. At the end of the withdrawal process, the quantity of product must match that identified initially; in case of deviations, these must be justified. The organization must define a frequency for market withdrawals simulation in order to verify that the procedure is maintained operational and effective. Although ISO 22000:2005 and its implementation guide does not provide a minimum period to carry out simulations, it is advisable that these are held annually (except in years in which effective withdrawals have been carried out). In fact, this is also the indication present in references BRC Issue 7, IFS Food version 6 and SQF Code edition 7.2.

## 4.6 Validation, verification, and improvement of food safety management system (Clause 8)

For relevant keywords, please refer to Figure 4.18.

**Figure 4.18** Keywords from Section 4.6.

### 4.6.1 General (Clause 8.1)

Clause 8 of the standard establishes the requirements that the organization should meet to validate, monitor, verify, and improve the food safety management system in order to demonstrate that it provides the expected level of control.

The update and enhancement of the food safety management system should guarantee that the information that was the basis of its development remains current and is characterized by a solid scientific basis originated from credible sources (e.g., academic, official bodies, relevant international organizations). It is the responsibility of the Food Safety Team to plan and implement the necessary processes to comply with this clause.

### 4.6.2 Validation of control measure combinations (Clause 8.2)

Despite this clause not having a direct correspondence to any of the stages of the HACCP methodology, in step 11 (*establishment of procedures of verification*) it is stated that, whenever possible, validation activities should be developed to confirm the effectiveness of all HACCP system elements.

ISO 22000:2005 defines the term 'validation' as the achievement of evidence that the control measures managed by the HACCP plan and by the OPRPs are effective (ISO 2005a). The validation focuses on gathering and assessing scientific, technical, and observational information, and should be performed prior to the implementation of the control measures or whenever changes are made to the control measures implemented. Validation of control measures should also be reassessed when new scientific or statutory information is produced or the products or process change. Figure 4.19 depicts the process of validation of the control measures (ISO 2014).

The definition of methods to demonstrate that the control measures, or combinations thereof, are able to control the risk in order to achieve the desired results should be (and commonly are) based on validated scientific/technical information

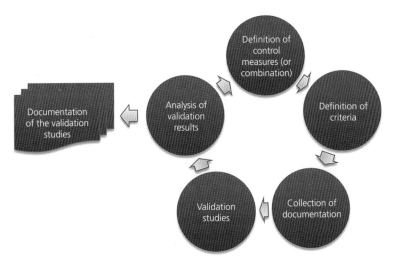

**Figure 4.19** Validation of the control measures.

and previous[20] or historical knowledge of the performance of control measures. Other methods suggested by ISO 22004:2014 are mathematical modeling and surveys. Validation studies should be preceded by the gathering of information necessary for such an assessment and should result in a report record that supports the decision.

During the validation of the control measures, if the results are not satisfactory the control measures should be revised and improved. When this is necessary, the Food Safety Team may take different approaches that can include changing (ISO 2005a): control measures (e.g., in the process parameters, the level of accuracy and/or their combination); raw materials; manufacturing technologies; or the characteristics or intended use of the finished product.

### 4.6.3 Control of monitoring and measuring (Clause 8.3)

Equipment and measurement methods must be controlled by people with competence to ensure that these provide valid results. It is up to the organization to establish which monitoring and measurements should be carried out and which equipment is necessary to ensure that those procedures are implemented, and provide evidence of conformity with the established requirements (ISO 9001:2008b).

---

[20] When validation is based on past experience of the Food Safety Team or when it was carried out by other people, it must be demonstrated that the same conditions of the intended application remain valid (ISO 2014).

The standard identifies procedures and tasks that must be established whenever it is necessary to ensure the validity of results obtained by equipment and methods of measuring and monitoring (ISO 2005a):

- ensure the calibration or verification at specific intervals and whenever necessary proceed to their adjustment;
- identify all equipment[21] and protect it from damage and deterioration; and
- safeguard equipment from adjustments that might invalidate the measurement result (e.g., adjustments by noncompetent people).

The organization must guarantee that the accuracy of the equipment is appropriate for the measures performed. The uncertainty defined by the calibration must be taken into consideration when the equipment assesses critical limits (e.g., when defining temperature set point).

Noncalibrated equipment must be immediately identified and, when possible, physically segregated from the rest of the equipment in order to ensure that it will not be used until recalibration. The validity of the previous measuring results must be evaluated, as well as the consequences to the safety of food products. The use of software for measurement/monitoring requires prior validation of its ability to control the specific requirements.

### 4.6.4 Food safety management system verification (Clause 8.4)

Verification activities and methods are not the clearest subjects when implementing a FSMS. In fact, it was not one of the three original principles of HACCP but, as described in Box 4.12, it was soon found to be decisive for the success of HACCP. Verification is defined in ISO 9000:2005 as confirmation, through the provision of objective evidence, that specified requirements have been fulfilled. The entire FSMS should be subject to verification activities that may include (ISO 2014):

- review of records and documents (e.g., verify if the CCP or OPRP records have been completed correctly);
- evaluate whether a PRP or a process has operated within requirements (e.g., verify if hand cleaning is being correctly executed and efficient);
- verify if the training, calibration, maintenance, or cleaning plans are being correctly carried out;
- confirm that external documentation (e.g., regulatory/statutory and customer requirements, customer complaints, external audits) are being evaluated and used to improve the FSMS effectiveness; or
- end-product testing.

---

[21] Identification should state when the equipment was calibrated or be used to identify the equipment in the list of equipment (see Section 4.5.1, Prerequisite 5) where this information should be included.

**Box 4.12** The verification principle

> In Box 4.3 Dr William Sperber describes how in 1972 in Pillsbury they found that, despite the 'hole being monitored,' nothing was done to correct it. The story finishes below describing how the HACCP principles become 7 (from the original 3) in the subsequent years.
>
> Sure enough, the sifter had indeed been inspected regularly and the inspector had first noted 'hole in sifter screen' (see Box 4.3). Immediately we created another HACCP principle: take corrective actions when deviations occur at a CCP. In the same time period our engineers had established another CCP: establish critical limits at each CCP.
>
> By 1975 Pillsbury's HACCP system consisted of 5 principles. Principles 6 and 7, for verification and record-keeping, were added in 1992 by the US National Advisory Committee on Microbiological Criteria for Food and by the UN Codex Alimentary Commission. More than 20 years later this program with 7 principles is still in use worldwide. It has greatly assisted global food producers as it is a standard for food safety management that can be used in all countries that are signatories to the World Trade Organization.

## Internal audit

ISO 19011:2011 (ISO 2011) defines audit as a systematic, independent, and documented process for obtaining audit evidence and evaluating it objectively to determine the extent to which the audit criteria are met.

The audit is a key factor in assessing the effectiveness of the food safety management system and verification of compliance of the planned provisions with the requirements of ISO 22000:2005. The organization shall establish a procedure to define the responsibilities, the qualification, and appointment criteria of auditors. The evidence of competence of auditors should be documented and archived (e.g., *Curriculum vitae*, certificates of qualification). Internal audits should be conducted by auditors that ensure the independence and impartiality of the process and should be conducted by people external to the process/audited area or even to the the organization when necessary.[22]

Since the frequency of assessment of the FSMS is not defined by the standard, the organization should set time intervals to perform internal audits. These intervals are defined according to the importance of the processes and audited areas, as well as the results of previous audits. The processes/areas of greatest relevance to food safety or that have obtained poor results in previous audits shall be subject to more constant checks. The criteria, scope, frequency, and methods of audits should be defined (ISO 2005a).

---

[22] The guide for application of the standard (ISO 22004:2014) identifies the possibility, particularly in small businesses (one or two managers), of failure to meet these requirements fully. In these cases it is required that the manager that performs the functions of auditor is objective and ensures fairness in the process. Another alternative proposal is to establish a partnership with another small company so that their managers perform the internal audit of the other company.

Records from internal audits should be maintained and results communicated not only to the process/area managers but also to the Food Safety Team and, if necessary, to the top management. The noncompliances detected during the audits shall be treated in accordance with the requirements specified in Section 4.5.10.

### Evaluation of individual verification results

The results of verification activities should be registered and include information on (ISO 2014):

- the effectiveness of the food safety management system;
- the personnel responsible for its management and updating;
- the records associated with the monitoring activities and equipment calibration; and
- the results of the review of records and analyzed samples.

The Food Safety Team should regularly evaluate the results from the planned verification activities (Section 4.5.8). When this assessment does not demonstrate compliance with the requirements, the organization should act in order to obtain the required compliance. For this purpose, any requirements for revision of existing procedures and communication channels, the hazard analysis, the PRPs, OPRPs, the HACCP plan, and the effectiveness of human resource management and training activities (ISO 2005a) should be identified. All the activities that the organization considers necessary to perform in order to restore compliance should be recorded as evidence.

### Analysis of results of verification activities

The results of verification activities, including internal and external audits (conducted according to Section 4.6.4, 'Internal audit'), should be analyzed with the purpose of (ISO 2005a):

- assessing the system's global performance;
- identifying the need for updates and improvements;
- identifying tendencies that can compromise the safety of the products;
- gathering the necessary information for the planning of audits; and
- providing evidence of the effectiveness of the corrections and corrective actions performed.

The results of the analysis and resulting activities should be registered and communicated in appropriate form to the top management, namely as an input during the management review (Section 4.3.8), and should be used for the update of the FSMS.

## 4.6.5 Improvement (Clause 8.5)

ISO 22000:2005 determines that organizations seek to continuously improve and update the food safety management system. Figure 4.20 depicts situations where improvement and updating are usually applied (ISO 2014).

**Figure 4.20** Situations where improvement and updating are commonly applied.

## Continual improvement

The top management should ensure that the organization continually improves the efficiency of the food safety management system through activities listed in Table 4.8 (ISO 2005a).

The improvement activities should be carried out using the plan-do-check-act (PDCA) methodology as described in standard ISO 9001 for the improvement of the quality management system performance (ISO 2008b).

Top management should encourage a culture of continuous improvement to achieve better performance, implementing a cycle of goals/control/recognition/reward. As highlighted in Section 4.3, top management attitude is very important to inspire personnel to have a better attitude towards the FSMS. Despite the definition of food safety objectives being a top management responsibility (and should be strategic for the organization), the importance of recognition and reward when such objectives are achieved it is not always perceived. Without this approach, together with promoting the participation of all (even personnel that have no obvious relation to food safety), top management will find it very difficult to create an improvement culture in the organization.

**Table 4.8** Activities through which top management shall ensure the improvement of the system and examples of those activities

| Activities | Examples |
| --- | --- |
| Communication | Ensure that there is sufficient external information available to update the FSMS. Guarantee that issues that have an impact on food safety are communicated with personnel. |
| Management review | The output of the FSMS performance evaluation should include decisions for its improvement. New food safety objectives and updated food safety policy. |
| Internal audits | Results from internal audits shall be discussed in the management review or even force the management to take immediate action (corrections or corrective actions) related to the identification of nonconformities. |
| Evaluation of individual verification results | Review of the training plan or PRP(s) found necessary after results of the verification activities. |
| Analysis of results of verification activities | Take action after identifying a trend that can generate potentially unsafe products. |
| Validation of the combinations of control measures | Change control measures or define new combinations when validation fails to prove its effectiveness. |
| Corrective actions | Take actions to eliminate the cause of a nonconformity and guarantee that the problem is not repeated in the future. |

## Updating the food safety management system

The food safety management system should be regularly evaluated and updated. It is the responsibility of the Food Safety Team to set the time intervals between each assessment according to the needs identified for the review of hazard analysis, the established operational PRPs, and the HACCP plan.

Any activity of updating should be recorded (and utilized in the management review) and, according to the guide on the application of ISO 22000 (ISO 22004:2014), may result from:

- new information from internal and external communication;
- results of the evaluation of the efficiency and effectiveness of the food safety management system (e.g., trend analysis, number of nonconformities, customer complaints, observations of daily operations);
- results of verification activities (e.g., internal/external audit); and
- guidelines resulting from management review (e.g., resource requirements, changes in the food safety policy and objectives).

# CHAPTER 5

# The FSSC 22000 certification

## 5.1 History

The recent history of the FSSC 22000 scheme and its main landmarks are depicted chronologically in Figure 5.1. The Foundation for Food Safety Certification (FFSC) developed and has the legal ownership of the scheme. Founded in 2004 by a group of Dutch certification organizations, the FFSC is a nonprofit organization whose goal is to maintain and act as a legal basis for the Dutch HACCP, governed by strict bylaws to ensure its independency (FFSC 2015c).

In October 2008 PAS 220:2008 was released, sponsored by Danone, Kraft Foods, Nestlé, and Unilever, through the Confederation of the Food and Drink Industries of the European Union (CIAA), later (June 2011) renamed to FoodDrinkEurope (BSI 2008b). These big manufacturers were aiming for 'a common set of prerequisite programs that can be used by any manufacturer who wishes to establish an ISO 22000 certified food safety management system.' Steve Mould, the technical author of PAS 220, also emphasized that this standard was created as a result of joint effort in the industry, after a wide consulting and public review conducted by the British Standards Institution (BSI 2008a).

Soon after PAS was released, the CIAA invited the FFSC to develop a food safety management scheme based on ISO 22000:2005 but using PAS 220:2008 to establish a specific prerequisite program and introduce the necessary additional requirements to achieve GFSI recognition. The experience of FFSC with Dutch HACCP (similar approach as ISO 22000), and the fact that it is a nonprofit organization and independent of retailers, was instrumental in accepting this invitation. Another key aspect that distinguished ISO 22000 from other food safety systems that were already GFSI recognized (BRC since 2000, IFS since 2002, SQF since 2004) at the time was that it is a management system certification standard and

*Food Safety in the Seafood Industry: A Practical Guide for ISO 22000 and FSSC 22000 Implementation,*
First Edition. Nuno F. Soares, Cristina M. A. Martins and António A. Vicente.
© 2016 John Wiley & Sons, Ltd. Published 2016 by John Wiley & Sons, Ltd.

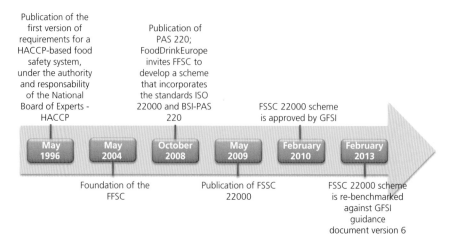

**Figure 5.1** Chronological events of the FSSC 22000 history.

not a process/product certification scheme. This new scheme (FSSC 22000) would have therefore a stronger focus on management commitment and continuous improvement, a more profound audit approach, and an easier integration with other ISO standards such as ISO 9001 or ISO 14001.

Since the new scheme was based on two previously published documents (ISO 22000:2005 and PAS 220:2008), the main challenges were to set additional requirements, aiming not only at giving a broader perspective to some ISO 22000 technical issues, but also to introduce new issues to comply with GFSI requisites. The requirements of GFSI for the governance of the scheme implied an effort to adopt a structure that would own and manage the scheme, while also guaranteeing a complete integrity program and monitoring certified organizations.

## 5.2 Scope

The FFSC 22000 scope is established using as reference the ISO/TS 22003 for the definition of the food chain categories to which the requisites apply. The last version of this technical specification (published in 2013) has only 11 categories, of which FSSC scope includes: C (food manufacturing); D (animal feed production); K (production of (Bio)chemicals); and I (production of food packaging and packaging material). In the future these categories will be extended to other food chain sectors as mentioned by Fons Schmid (Chairman of the Board of Stakeholders of FSSC 22000) in a Lloyd's Register Quality Assurance (LQRA) webinar (October 2013). He emphasized that the FSSC Board of Stakeholders (BoS) takes requests from the market to extend the FSSC scope of certification very seriously. Nevertheless, storage, warehousing, and distribution may be added to a manufacturing scope when it is dedicated to the company's own production, included within the audited safety management system and a part of the same legal entity (FFSC 2014). If not, these activities can only be assessed through ISO 22000 standard. In February 2015, the FFSC published the 3.2 version of the scheme to include animal production and a voluntary quality module based on ISO 9001.

### 5.2.1 Certification scope

As mentioned in Section 4.2.1 it is necessary to define the *certification scope*, which is the responsibility of the Certification Bodies (CBs). This definition assumes great importance; as well as being printed in the organization certificate, it sets the framework for the audit. This is why the certification scope is commonly discussed and agreed between the CB and the company that is applying for certification, eventually even before the first visit of the CB auditors or in a preliminary audit. Some of the most important requirements that are mandatory for the certification scope definition are (FFSC 2014):

1. not to be misleading or vague;
2. describe the main activities, audited process, and products; and
3. not hold promotional statements.

The FSSC 22000 guidance document on certification scopes gives some examples of correct and incorrect scope descriptions (FFSC 2014). Another practical way to analyze and study examples of certification scopes is by accessing the FSSC 22000 online directory of the certified organizations (FFSC 2015a). A large variety of certification scopes can be found, examples of which are given below. Examples 1 and 2 are very detailed, containing references to processes, products, and even packaging type, while examples 3–5 include other activities such as warehousing, transport, and storage.

1. The sorting, gibbing or deheading, and salting of herring, filleting of mackerel, and re-hydrating of stockfish. Packaging of (fresh and frozen) fish and fish products into cartons, buckets, and film (vacuum bags): Werner Larsson Fiskeeksport A/S.
2. Freezing, glazing, salting, injecting, sorting, packing, repacking, portioning, cutting, and breaking of fish: Quick Frozen B.V.
3. Purchase, processing, packing, and supply of fresh mussels and oysters and the processing, packing, storage, and supply of cooked preserved shellfish (frozen, chilled, and ambient stable): Prins Groep B.V.
4. Processing, warehousing, and sale of chilled and frozen fishing products: Mobilpesca Surgelati S.p.A.
5. Processing, preservation, storage, and transportation of fish and fish products: Pescado Grup S.R.L.

## 5.3 Prerequisite program

The prerequisite program has been discussed in detail in Section 4.5.2. The novelty brought by FSSC 22000 to ISO 22000 is the mandatory use of technical specifications for the sector (in addition to Clause 7.2 of ISO 22000) which, in the case of the seafood industry, is ISO/TS 22002-1: *Prerequisite programs on food safety – Food manufacturing* (FFSC 2015b). Other sources of information are also identified and shall be considered, such as regulatory requirements, recognized codes of practice and guidelines, and customer requirements. From these sources, if regulatory requirements are clear and easy to find in the legislation of the country where the organization is established, it can be more complicated to identify which are the 'recognized' codes or guidelines applicable. Information on customer requirements is even trickier because what future customers may request cannot be anticipated; further, if such requests are related to layouts and facilities, these can be quite expensive and difficult to change. This case is particularly important for organizations that export their products to different countries.

Finally, it is explicitly mentioned that the organization can exclude or find alternatives to a prerequisite as long as it still complies with the original prerequisite requirements and is well documented and supported by a hazard analysis.

Table 5.1 clarifies the correspondence between ISO/TS 22002-1:2009 and the prerequisites of ISO 22000:2005 (present in Clause 7.2 of the standard) and the General Principles of Food Hygiene (CAC 1969).

Although ISO/TS 22002-1:2009 is the most recent document, only two of its clauses cannot be found in the structure of the General Principles of Food Hygiene of *Codex Alimentarius* (CAC/RCP 1-1969). When compared to ISO 22000:2005 (Clause 7.2), ISO/TS 22002-1:2009 clauses overlap completely from 1–10 inclusively (corresponding to subparagraphs a) to j)). Despite Clauses 12, 13 and 14 not having a direct correspondence, those subjects are discussed in other clauses of the standard as referred to in the table.

The mandatory use of the prerequisite technical specifications does not introduce a radical change to already established prerequisite programs based on ISO 22000:2005 or *Codex Alimentarius*. However, it is a document that develops in detail each of the prerequisites, has been specifically targeted at food manufacturing organizations, and contains effective guidelines regarding how to deal with reprocessed products and defend itself from deliberate attacks of its products.

## 5.4 Additional requirements

### 5.4.1 Specifications for services

This requirement aims at reinforcing the aspects related to control of support services suppliers such as utilities, transportation, and maintenance. There are references to the need for this control in General Principles of Food Hygiene (Clauses 5.3, 5.4, and 5.5) where specific requisites regarding raw materials, ingredients, water, and packaging management are mentioned. In the prerequisites of ISO 22000:2005 (Clause 7.2.3.f), the management of purchased products is generalized regarding the provision of other services, for example water and its derivatives, waste management, storage, and transport.

**Table 5.1** Correspondence between the prerequisites of ISO/TS 22002-1:2009 and ISO 22000:2005 and the general principles of food hygiene of *Codex Alimentarius*

| ISO/TS 22002-1:2009 | ISO 22000:2005 | CAC/RCP 1-1969 |
|---|---|---|
| 1. Construction and layout of buildings (Clause 4) | a) Construction and lay-out of buildings and associated utilities | Section IV: Establishment: Design and facilities |
| 2. Layout of premises and workspace (Clause 5) | b) Lay-out of premises, including workspace and employee facilities | Section IV: Establishment: Design and facilities |
| 3. Utilities: air, water, and energy (Clause 6) | c) Supplies of air, water, energy, and other utilities | Section V: Control of operation; 5.5 Water |
| 4. Waste disposal (Clause 7) | d) Supporting services, including waste and sewage disposal | Section VI: Establishment: Maintenance and sanitation; 6.4 Waste management |
| 5. Equipment suitability, cleaning, and maintenance (Clause 8) | e) Suitability of equipment and its accessibility for cleaning, maintenance, and preventative maintenance | Section VI: Establishment: Maintenance and sanitation; 6.1 Maintenance and cleaning |
| 6. Management of purchased materials (Clause 9) | f) Management of purchased materials; disposals and handling of products | Section V: Control of operation; 5.3 Incoming material requirements; 5.4 Packaging |
| 7. Measures of prevention of cross-contamination (Clause 10) | g) Measures for the prevention of cross-contamination | Section V: Control of operation; 5.2.4 Microbiological cross-contamination |
| 8. Cleaning and sanitizing (Clause 11) | h) Cleaning and sanitizing | Section VI: Establishment: Maintenance and sanitation; 6.2 Cleaning programs |
| 9. Pest control (Clause 12) | i) Pest control | Section VI: Establishment: Maintenance and sanitation; 6.3 Pest control system |
| 10. Personnel hygiene and employee facilities (Clause 13) | j) Personnel hygiene | Section VII: Establishment: Personal hygiene |
| 11. Rework (Clause 14) | | |
| 12. Product recall procedures (Clause 15) | (see Clause 7.10.4) | Section V: Control of operation; 5.8 Recall procedures |
| 13. Warehousing (Clause 16) | (see Clause 7.2.3a) | Section IV: Establishment: Design and facilities; 4.4.8 Storage |
| 14. Product information/ consumer awareness (Clause 17) | (see Clause 5.6.1b) | Section IX: Product information and consumer awareness |
| 15. Food defense, biovigilance, and bioterrorism (Clause 18) | | |
| | | Section X: Training |
| | | Section VIII: Transportation |

This additional requirement defines that any provided service which may have an impact in food safety should: (1) have specified requirements for its control and use; (2) be described in documents, allowing assessment of its impact on the processes and in food safety; and (3) be managed in compliance with technical specifications defined for sector prerequisites (FFSC 2015b).

ISO/TS 22002-1 already envisaged the need to establish specific requirements for some services, for example:

- Water (Clause 6.2): 'Water used as a product ingredient, including ice and steam ... shall meet specified quality and microbiological requirements relevant to the product.'
- Air (Clause 6.4): 'The organization shall establish requirements for filtration, humidity (RH%) and microbiology of air used as an ingredient...'
- Waste (Clause 7.3): 'Removal frequencies shall be managed to avoid accumulations.'
- Maintenance (Clause 8.6): 'Corrective maintenance shall be carried out in such a way that production on adjoining lines or equipment is not at risk of contamination.'
- Pest control (Clause 12.2): 'Pest management programs shall be documented and shall identify target pests, and address plans, methods, schedules, control of procedures and, when necessary, training requirements.'

Other examples of services for which requirements may need to be defined include the following.

- Cleaning (uniforms): Define adequate cleaning agents, processes, and control parameters.
- Employees' health status evaluation: Define the frequency and kind of medical exams, as they could be different depending on the relevance of the employees' function to food safety.
- IT services: The growing use of external IT services to store information and documentation (in so-called cloud servers) means it is important for the organizations to establish procedures (together with the service supplier) to guarantee the safety and protection of the documentation necessary to support the food safety system (e.g., manuals, traceability, and other records).

This requirement also defines the need to describe services in detail in order to make a complete hazard analysis. Accordingly, the organization should identify aspects such as:

- the definition of responsibilities of recruitment, selection, validation of technical skills, and communication with the company providing the service;
- the stage of the process and products that may be affected;
- who will verify (and how) that the defined requirements are being complied with; and
- how the management of these services fulfills the requirements defined in the technical specifications.

Although this requirement presents Clause 7.3.3 of ISO 22000:2005 as a reference, which contains the necessary information to characterize the product, a parallel cannot be drawn with the information necessary to characterize services which will almost always be of a different nature. However, the reference made to the obligation of identifying the statutory and regulatory requirements and the need for specifications to be kept up to date are clearly also applicable to services.

## 5.4.2 Supervision of personnel in application of food safety principles

This requirement identifies the need for supervision of the personnel in the application of the food safety principles and practices. This clause offers a more general perspective than Clause 6.2.2 of ISO 22000:2005, which is used as reference. That clause highlights the importance of developing skills for 'personnel whose activities have an impact on food safety' and providing training for 'personnel responsible for monitoring, corrections and corrective actions.' This additional requirement corresponds best to what is expressed in the General Principles of Food Hygiene (CAC 1969), which state the importance of proper training and supervision of all employees involved in activities which could have an effect on food safety. Clause 10.3 of the same document defines the need to establish supervision and verification routines to ensure that procedures are being met effectively.

To comply with the principle set out in this requirement, the supervision function cannot be too centralized. This situation is sometimes common, especially in SMEs and small companies, where there is a large concentration of knowledge and responsibility in the quality/food safety manager. It would therefore be advantageous to equip at least one person from each working group, shift, or department with a deeper understanding of the principles and procedures of food safety to guarantee supervision at all locations, moments, and activities related to food products.

## 5.4.3 Specific regulatory requirements

As stated in appendix IA of FSSC 22000 Guidance, this requirement is an extension of Clause 7.3.3.1 of ISO 22000:2005. This clause presents a list of topics that organizations must describe for each raw material, ingredient, and product-contacting material in order to perform the hazard analysis. The identification of all statutory and regulatory food safety requirements is also mandatory in this description.

What FSSC 22000 promotes with this additional requirement is the extension of its application to all substances (not just those requiring hazard analysis) and all statutory and regulatory requirements (not just those related to food safety). The organization must have a broader knowledge about the materials and ingredients they use to develop more extensive and complete specifications and control.

Appendix IA of FSSC 22000 Guidance provides examples of the type of requirements to consider when preparing the specifications, including prohibited colors and maximum preservative levels. These two examples are of particular interest to the fish industry, particularly when organizations use products from aquaculture (whose composition must be controlled due to the use of feeds) or crustaceans and cephalopods (in which case substances such as phosphates are currently used to preserve the product and retain water); see Box 5.1.

**Box 5.1**   Phosphates in the seafood industry

The use of phosphates in seafood is controversial both in terms of the justification for its use and the determination of their concentration in the product. The use of phosphates is usually justified by the improvements in the retention of natural moisture by proteins (reducing the amount of thaw drip), inhibition of flavor degradation and lipid oxidation, aiding emulsification and removal of shell fish shells, and offering cryoprotection (Seafish 2012). As a consequence however, the ability of the product to retain added water can promote abusive situations and fraud. The fact that phosphates are a naturally occurring substance imparts a greater complexity to the control of its usage because it requires an analytical method capable of distinguishing between natural and added phosphates. In the Report of the 33rd session of the Codex Committee of Fish and Fishery Products, held in February 2014, changes to the maximum phosphates quantities that may be present in certain fish and fishery products were analyzed and proposed, which is a clear sign of the relevance of these two issues. In some cases, the use of phosphate as an additive was not considered to be technologically justifiable. In cases where a threshold is set, it was discussed whether or not this value should include the naturally occurring content in the product.

The amendments proposed by this committee were adopted at Meeting No. 37 of the Committee of the *Codex Alimentarius*, held in July 2014. Nevertheless, there were disagreements among its members; for example, the delegation of Brazil did not consider it possible to identify frauds if the set limit values included natural phosphate and the delegation of Nigeria did not consider the use of phosphates in the products that were being assessed justifiable.

## 5.4.4 Announced but unscheduled audits of certified organizations

This additional requirement is intended for certification bodies. They are required to participate in an audit program which includes the monitoring of audits conducted to certified companies. As described in appendix IA of the FSSC 22000 Guidance, the Foundation will announce to the CBs the number of audits that will be supervised at the beginning of the year, but the locations and dates will be chosen from the plan of the CBs. This additional requirement is in response to that required by the GFSI on this subject, and is not related to unannounced audits to certified organizations described in Box 5.2.

## 5.4.5 Management of inputs

The guidance for application of additional requirements emphasizes that the focus of this clause are the inputs necessary to the confirmation of product safety and not the inputs to achieve or maintain it.

This additional requirement has two main challenges:

1. definition of the inputs that are critical to the confirmation of product safety; and
2. use of an accredited testing protocol or definition of a test protocol that is equivalent to those described in ISO 17025 to analyze the critical inputs.

Materials that are added to the product or used as processing aids that can potentially introduce contamination or change product properties, such as pH or moisture content, are used as examples of critical inputs (FFSC 2013).

**Box 5.2**   Unannounced audits

> The use of unannounced audits seems to be a growing trend, aiming to create greater confidence in the audit process. In fact, it is quite common for the scheduled audit day not to be a regular operations day of the audited organization. This subject is a concern for the organization itself when, for example, it is auditing suppliers.
>
> How three of the GFSI-benchmarked schemes address the possibility of audited schemes are described below. The GFSI scheme management subworking group is currently discussing the subject of unannounced audits. It is expected that this theme will be addressed in Guidance Document Version 7 (to be released in 2015/2016) and therefore all benchmarked schemes, including FSSC 22000, will have to comply with the new requirements.
>
> - *SQF Ed. 7.2*: One unannounced re-certification audit is mandatory within three certification cycles. The year when the unannounced audit will take place is determined between the organization and the certification body, and must be conducted around 30 days from the re-certification date.
> - *IFS Food Check*: Voluntary program ordered directly to IFS Management (that also assigns the auditors) which consists of an announced audit using a checklist based on criteria from IFS Food Standard. This certification body will then supplement the company's certificate with the comment 'The company takes part in the 'unannounced IFS Food Check' program' (IFS 2014).
> - *BRC Issue 7*: Unannounced program is voluntary. There are two options: single unannounced visit to audit the whole standard; or two separate visits, the first unannounced to audit the good manufacturing practices and the second planned and for mainly checking documents and records. The unannounced audit typically occurs in the last 4 months of the certification cycle and substitutes the scheduled audit. To the audit grade is added a '+' (e.g., A+, B+, or C+) and to the certificate the expression 'Unannounced Option'.

Regarding the use of an accredited testing protocol, it would be straightforward if there was always (and everywhere) available an ISO 17021 accredited laboratory with the intended test included in its accreditation scope. However, that is clearly not always the case. This additional requirement therefore allows organizations to perform tests in a laboratory or other testing facility, as long as it is demonstrated that the test protocol meets the principles of ISO 17025.

# Appendix 1

**Table A1.1** List of the documentation identified by ISO 22004:2014 and respective clauses of ISO 22000:2005

| Documentation | Clause no. | Clause name |
|---|---|---|
| **Documents used to document the FSMS** | | |
| Food safety policy | 4.2.1 | Food safety management system: Documentation requirements |
| List of measurable food-safety-related objectives | 4.2.1 | Food safety management system: Documentation requirements |
| Food safety requirements from statutory/regulatory authorities | 5.6.1 | Management responsibility: Communication |
| Procedures and instructions related to the PRP(s) includes monitoring procedures, where applicable, and verification plans | 7.2.3 | Planning and realization of safe products: Prerequisite programs |
| Competencies of the food safety team (e.g., curriculum vitae) | 7.3.2 | Planning and realization of safe products: Preliminary steps to enable hazard analysis |
| Description of raw materials, ingredients, and product-contact materials | 7.3.3.1 | Planning and realization of safe products: Preliminary steps to enable hazard analysis: Product characteristics |
| Description of the characteristics of end-products | 7.3.3.2 | Planning and realization of safe products: Preliminary steps to enable hazard analysis: Product characteristics |
| Flow diagrams | 7.3.5.1 | Planning and realization of safe products: Preliminary steps to enable hazard analysis: Flow diagrams, process steps, and control measures |

*Food Safety in the Seafood Industry: A Practical Guide for ISO 22000 and FSSC 22000 Implementation*, First Edition. Nuno F. Soares, Cristina M. A. Martins and António A. Vicente. © 2016 John Wiley & Sons, Ltd. Published 2016 by John Wiley & Sons, Ltd.

**Table A1.1** (Continued)

| Documentation | Clause no. | Clause name |
|---|---|---|
| Description of control measures, process parameters, and/or the rigorousness with which they are applied, or procedures that may influence food safety | 7.3.5.2 | Planning and realization of safe products: Preliminary steps to enable hazard analysis: Flow diagrams, process steps, and control measures |
| Description of external requirements that may impact the choice and the rigorousness of the control measures | 7.3.5.2 | Planning and realization of safe products: Preliminary steps to enable hazard analysis: Flow diagrams, process steps, and control measures |
| OPRPs | 7.5 | Planning and realization of safe products: |
| HACCP plan | 7.6.1 | Establishing the HACCP plan |
| – Including monitoring system | 7.6.4 | |
| – Including corrective action | 7.6.5 | |
| Verification plan | 7.8 | Planning and realization of safe products |
| Rationale and supporting documentation for chosen critical limits | 7.6.3 | Planning and realization of safe products: Establishing the HACCP plan |
| List of documents applicable to the FSMS | Optional | |
| Organizational charts | Optional | |
| List of members of the food safety team (internal and external experts) | Optional | |
| Layout of premises and buildings | Optional | |
| Job descriptions | Optional | |
| Contractual agreements and/or warranty statement from subcontractors | Optional | |
| List of external reference documents consulted when establishing PRPs | Optional | |
| Composition, organization, and contact details of the food safety team responsible for addressing possible emergency situations and potential accidents should be documented and kept up to date | Optional | |

**Documents that describe how to carry out an activity and/or intended input or output**

*Documented procedures*

| | | |
|---|---|---|
| Procedure for the control of documents | 4.2.2 | Food safety management system: Documentation requirements |
| Procedure for the control of records | 4.2.3 | Food safety management system: Documentation requirements |
| Procedure for corrections | 7.10.1 | Planning and realization of safe products: Control of nonconformity |
| Procedure for corrective actions | 7.10.2 | Planning and realization of safe products: Control of nonconformity |

**Table A1.1** (Continued)

| Documentation | Clause no. | Clause name |
|---|---|---|
| Procedure for handling of potential unsafe products | 7.10.3 | Planning and realization of safe products: Control of nonconformity |
| Withdrawal/recall procedure | 7.10.4 | Planning and realization of safe products: Control of nonconformity |
| An internal audit procedure | 8.4.1 | Validation, verification, and improvement of the food safety management system: Food safety management system verification |
| Procedures for identifying product/ batch codes and traceability | Optional | |
| Procedures for internal and external communication | Optional | |
| *Instructions that are required by the International Standard* | | |
| Competencies can be ensured through education, training, or by other appropriate means | 7.6.3 | Planning and realization of safe products: Establishing the HACCP plan |
| Protocol for saving computer data, where documentation is kept electronically | Optional | |
| Instructions required to implement and maintain PRPs | Optional | |
| Manufacturer's instructions for use of detergents and sanitizers | Optional | |
| Instructions for transportation | Optional | |
| Instructions for staff hygiene | Optional | |
| *Specifications* | | |
| Specifications describing acceptable levels of hazards for ingredients, including additives and processing aids | Optional | |
| Specifications describing the acceptable levels of hazards for end products | Optional | |
| Specifications for contracted/ outsourced services | Optional | |
| Specifications for methods/conditions of production | Optional | |
| Specifications for physical parameters | Optional | |
| Specifications for food contact material | Optional | |
| Water quality requirements | Optional | |
| Compressed gas specifications | Optional | |
| **Records** | | |
| Information concerning food safety obtained through external communication | 5.6.1 | Management responsibility: Communication |

(Continued)

**Table A1.1** (Continued)

| Documentation | Clause no. | Clause name |
|---|---|---|
| Information about changes that may impact food safety | 5.6.2 | Management responsibility: Communication |
| Input to and output of management reviews | 5.8 | Management responsibility |
| Agreements or contracts defining the responsibility and authority of external experts or contractors assisting the food safety team or supporting the FSMS | 6.2.1 | Resource management: Human resources |
| Training or other actions to ensure personnel competencies | 6.2.2 | Resource management: Human resources |
| Changes to PRPs | 7.2.3 | Planning and realization of safe products: Prerequisite programs |
| Flow diagrams | 7.3.5.1 | Planning and realization of safe products: Preliminary steps to enable hazard analysis: Flow diagrams, process steps, and control measures |
| Food safety hazards that have been identified as reasonably expected to occur | 7.4.2.1 | Planning and realization of safe products: Hazard analysis: Hazard identification and determination of acceptable levels |
| Justification for, and the result of the determination, regarding the acceptable level of a food safety hazard assessment | 7.4.3 | Planning and realization of safe products: Hazard analysis |
| The methodology parameters used and results for the categorization of control measures | 7.4.4 | Planning and realization of safe products: Hazard analysis |
| Monitoring records for OPRPs and CCPs | 7.5, 7.6.1 | Planning and realization of safe products: Establishing the HACCP plan |
| Results of verification | 7.8 | Planning and realization of safe products |
| Traceability records | 7.9 | Planning and realization of safe products |
| Results of evaluation of potentially unsafe products and of corrections | 7.10.1 | Planning and realization of safe products: Control of nonconformity |
| Corrective actions taken | 7.10.2 | Planning and realization of safe products: Control of nonconformity |
| The cause, extent, and result of a withdrawal/recall | 7.10.4 | Planning and realization of safe products: Control of nonconformity |
| Results of verification regarding the effectiveness of the withdrawal/recall program | 7.10.4 | Planning and realization of safe products: Control of nonconformity |
| Results of calibration and verification of measuring equipment | 8.3 | Validation, verification, and improvement of the food safety management system |
| Results of assessment regarding nonconforming measuring equipment, including actions on affected product | 8.3 | Validation, verification, and improvement of the food safety management system |

**Table A1.1** (Continued)

| Documentation | Clause no. | Clause name |
| --- | --- | --- |
| System updating activities | 8.5.2 | Validation, verification, and improvement of the food safety management system: Improvement |
| Implementation schedule of planned activities | Optional | |
| Communication plan | Optional | |
| Results of validation and the method used | Optional | |
| Results of verification for outsourced or subcontracted activities | Optional | |
| Records of each withdrawal/recall event | Optional | |

# Appendix 2

## Introduction

All data that have served as a basis for the analysis presented in this Appendix was obtained from a database created by the European Commission in order to make the information on hazard occurrences available to consumers, organizations, and authorities from all over the world (RASFF: Food and feed safety alerts). Created in 1979, it allows food safety risks notifications to be shared with the aim of preventing such foods from reaching consumers.

The analysis is based on research conducted in May 2015 through the official webpage of the RASFF.[1] There are several criteria available to perform the search. The parameters which were selected to be used during this study were:

- *product category*: bivalve mollusks and products thereof; cephalopods and products thereof; crustaceans and products thereof; and fish and fish products;
- *hazard category*: allergens; biocontaminants; biotoxins; food additives; foreign bodies; heavy metals; industrial contaminants; migration; parasitic infestation; pathogenic microorganisms; pesticide residues; and residues of veterinary medicinal products; and
- *date*: until 31 December 2014.

No other criteria (e.g., notification, type, keywords) were used to narrow the research. After the application of the criteria listed above and the compilation of all results in a table, a list with 5761 notifications was obtained. Table A2.1 presents the first five occurrences of that list.

---

[1] http://ec.europa.eu/food/safety/rasff/index_en.htm

---

*Food Safety in the Seafood Industry: A Practical Guide for ISO 22000 and FSSC 22000 Implementation*,
First Edition. Nuno F. Soares, Cristina M. A. Martins and António A. Vicente.
© 2016 John Wiley & Sons, Ltd. Published 2016 by John Wiley & Sons, Ltd.

**Table A2.1** Example of information obtained from RASFF (O: country of origin)

| Date | Ref. | Product type | Notification Type | Basis | By | Countries concerned | Subject | Action taken | Distr. status |
|---|---|---|---|---|---|---|---|---|---|
| 10.04.1980 | 1980.03 | Food | Alert | | Belgium | Belgium, Malaysia (O) | Staphylococcus in shrimps, frozen | | |
| 14.10.1981 | 1981.07 | Food | Alert | | France | Cuba (O), France | Toxin unknown in lobster, frozen/cooked | | |
| 25.11.1981 | 1981.09 | Food | Alert | | United Kingdom | Morocco (O), United Kingdom | Histamine in sardines | | |
| 07.02.1982 | 1982.02 | Food | Alert | | Denmark | Denmark, United States (O) | Botulism in salmon, canned | | |
| 09.03.1983 | 1983.06 | Food | Alert | | Denmark | Denmark, Japan (O) | Mercury in shark, canned/ smoked | | |

In order to perform the intended analysis, it was necessary to introduce new criteria. The following information was therefore added to each notification:

- *product category*: identifies the product category for each notification according to the list mentioned above;
- *hazard category*: identifies the hazard category for each notification according to the list mentioned above;
- *hazard classification*: classifies each notification as chemical, physical, or biological hazard (according to Section 1.3);
- *hazard*: identifies the contaminant within each hazard category (obtained from the information contained in 'subject'; see Table A2.1);
- *food*: identifies each fish or fish product notified (obtained from the information contained in 'subject'; see Table A2.1); and
- *physical state (packaging)*: identifies the preservation state or packaged state declared in the notifications (obtained from the information contained in 'subject'; see Table A2.1).

Regarding the information collected from the criterion 'subject' (hazard, food, and physical state) it was possible to verify that in some cases it was missing. In these cases, the notification was classified as non-declared (ND) and it was excluded from the criteria under consideration.

On the other hand, it was extremely difficult to homogenize the information, especially in the criterion 'physical state.'[2] In some situations, the identified information was represented by words or expressions with very similar meaning. This problem was addressed to Europe Direct Contact Centre, who explained:

> … you need to take into account the wide source of information that RASFF received, from all the members of the network, the urgency to circulate the information, and the variety of languages used in the involved labeling of the products. The RASFF team checks the information mentioned on all the notifications, verifying the accuracy and the completeness.

It was also mentioned that the problem is being addressed and that in 2014 the 'Standard operating procedures (SOP) of the Rapid Alert System for Food and Feed' was released, in an attempt to homogenize information (especially the content of the 'subject') reported by the authorities that file the format.

## RASFF notifications analysis

### Analysis of the notifications according to hazard classification

The analysis was conducted using a total of 5761 notifications distributed for all seafood categories. Analysis of Figure A2.1 reveals the following.

---

[2] An extensive variety of information is included in this criterion. For this study, only the 6 most common (chilled, frozen, live, fresh, canned, and smoked) were considered since the sum of the remaining 14 criteria only represented 2% of the notifications.

- In the first 17 years, the number of notifications was low (41, representing less than 0.7% of the total); however, this number increases significantly after 1997.
- Of all notifications, 64.1% originate from chemical hazards, 35.1% from biological hazards, and 0.8% from physical hazards.
- The year of the greatest numbers of notifications was 2009 (527).

In order to analyze the trend of each hazard classification, notifications were grouped in sets of 5 years. Only the 5577 notifications subsequent to 2000 were considered since, prior to this date, the number of notifications was low and did not include all hazard classifications.

Analysis of Figure A2.2 reveals the following.

- Biological hazards remained stable during the first two sets of 5 years analyzed (an average of 498 notifications); however, in the last period there was a significant increase (879 notifications).
- Chemical hazards present the greatest number of notifications, although in the last period of analysis there was a significant decrease (29%) from the previous period.
- Over the last two periods of analysis, the decrease of chemical hazard notifications was higher than the increase of biological hazards, which resulted in a reduction of notifications from a total of 2259 (between 2005 and 2009) to 2114 (between 2010 and 2014).

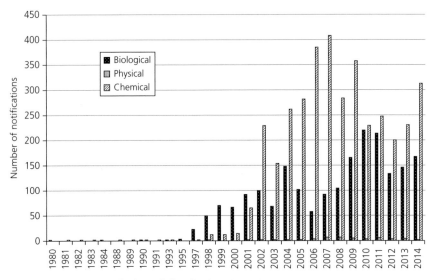

**Figure A2.1** Notifications by hazard classification since 1980.

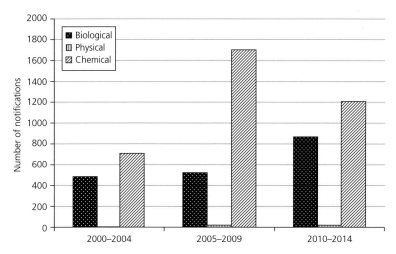

**Figure A2.2** Notifications by hazard classification grouped in sets of 5 years (2000–2014).

## Analysis of the notifications by hazard

This section of the study identifies and analyzes the main notified hazards in each hazard classification.

### Biological hazards

The five main hazards represent a total of 84.2% of occurrences; *Listeria monocytogenes* and *Anisakis* spp. were the most reported hazards. Figure A2.3 depicts the evolution of the notifications for each of the five main biological hazards identified in Table A2.2.

- *Vibrio* spp. reports have been decreasing in the last few years, dropping to a single notification in 2014; on the other hand, *E. coli* has shown a growth trend in notifications, although this was interrupted from 2011 to 2012.
- The number of reports for *E. coli* during this period is higher than that for *Vibrio* spp., which contradicts data presented in Table A2.2. In fact, this is a recent trend and was not observed before 2004. During that period, the number of notifications for *Vibrio* spp. (249) represented 81% of the total of reports for this hazard, whereas *E. coli* had only 3 reports.
- After a 6-year period (2006–2011) of constant increase, reports for *Anisakis* spp. were significantly lower over the last 3 years under analysis.

  Figures A2.4–A2.7 identify and evaluate which product and physical state (packaging) were reported for the two most reported biological hazards during 2005–2014. Analysis of Figure A2.4 reveals that most notifications for *Lysteria monocytogenes* are in salmon. Analysis of Figure A2.5 reveals the following two points.
- During the period considered the vast majority of notifications for *Listeria monocytogenes* were in smoked (51.1%), frozen (17.9%), and chilled (16.2%) products.

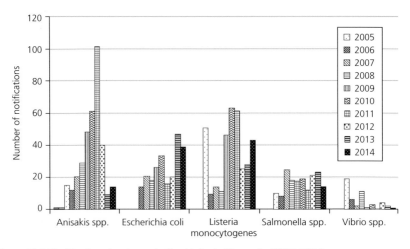

**Figure A2.3** Notifications for the main five biological hazards (2005–2014).

**Table A2.2** Main biological hazards notified since 1980

| Hazard | % |
| --- | --- |
| *Listeria monocytogenes* | 24.9 |
| *Anisakis* spp. | 19.5 |
| *Salmonella* spp. | 14.4 |
| *Vibrio* spp. | 13.7 |
| *Escherichia coli* | 11.7 |

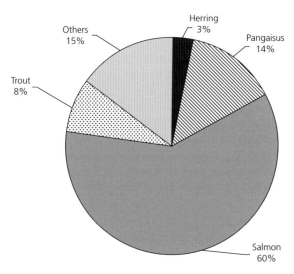

**Figure A2.4** *Listeria monocytogenes* notifications divided by product.

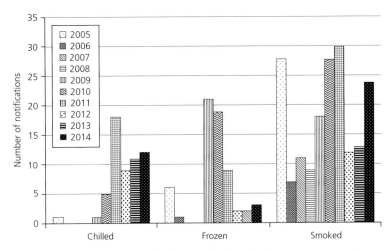

**Figure A2.5** *Listeria monocytogenes* notifications divided by physical state (packaging).

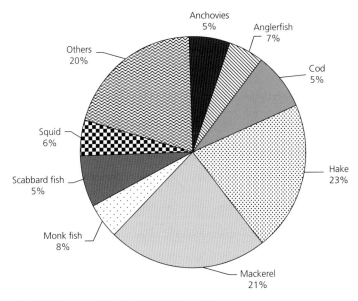

**Figure A2.6** *Anisakis* spp. notifications divided by product.

- Notifications in chilled products have been increasing since 2009, in contrast to that from frozen products.
  Analysis of Figure A2.6 reveals that:
- Notifications for mackerel and hake represent 44% of the total reports.
- In the 'Others' category 22 different species are included, with 1–2 notifications per species.

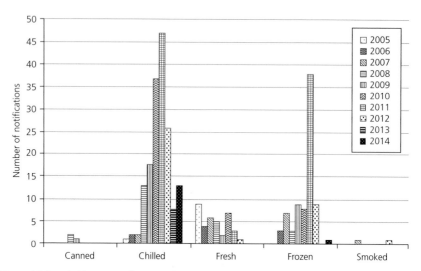

**Figure A2.7** *Anisakis* spp. notifications divided by physical state (packaging).

Analysis of Figure A2.7 reveals the following.

- It is not clear if there is any tendency in notifications for frozen and fresh state. After reaching its maximum number in 2001, notifications for chilled state were relatively low over the last 2 years of the analysis.
- The chilled state represents 48% of all reports, while fresh and frozen states represent 11% and 22%, respectively. The percentage of notifications in the chilled physical state (packaging) for this hazard (48%) is much higher than the percentage this state has in total (21%). This reveals that there is a higher preponderance of this parasite when the product is in this physical state.

## Chemical hazards

The number of chemical agents reported is very extensive (109) in the period. Due to this fact, it was decided to analyze only the six main hazards (benzo(a)pyrene, cadmium, histamine, mercury, nitrofuran, and sulfite) which represent 76% of all reports for chemical agents. Analysis of Figure A2.8 reveals the following.

- Mercury was the most reported hazard over the last 10 years, accounting for 37% of all notifications.
- The years with the highest number of notifications were 2007 and 2009 (319 reports each). On the other hand, 2012 was the year with the least number of notifications (156).
- Sulfites and nitrofurans present a consistent decrease in notifications. However, in the last year of the analysis, the number of reports for nitrofuran was twice the average of the previous 4 years.

Figures A2.9 and A2.10 identify and evaluate which product and physical state (packaging) were found in mercury reports.

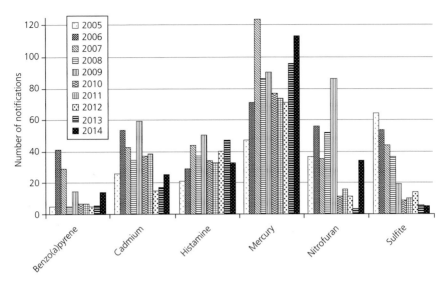

**Figure A2.8** Notifications for the six main chemical hazards (2005–2014).

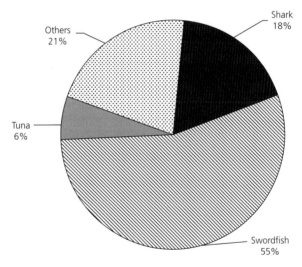

**Figure A2.9** Mercury notifications divided by products (2005–2014).

- According to Figure A2.9, there are two main products reported (73%) for mercury contamination: swordfish and shark.
- The most reported physical state (packaging) is frozen (Fig. A2.10), representing 45% of all notifications.
- There is an increasing trend in the number of reports for the chilled state (Fig. A2.10).

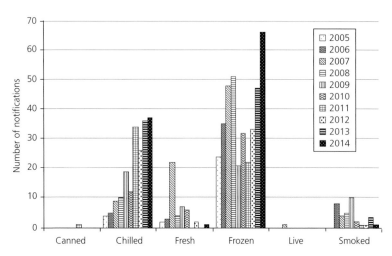

**Figure A2.10** Mercury notifications divided by physical state (packaging).

**Table A2.3** Main physical hazards notified (2005–2014)

| Hazard | No. notifications |
|---|---|
| Bone | 1 |
| Fish hook | 1 |
| Glass | 5 |
| Insects | 11 |
| Metal | 8 |
| Mineral | 1 |
| ND | 7 |
| Plastic | 5 |
| Rodent | 1 |
| Sand | 1 |

## Physical hazards

The number of physical hazards reported is very low. Further, 7 of the 41 notifications either did not mention the hazard or its identification was ambiguous (e.g., 'sharp'). Table A2.3 lists all hazards reported from 2005 to 2014. Prior to 2005, only seven reports were made. Of all physical hazards identified, four types represent 71% of all reports: glass, insects, metal, and plastic.

## Analysis of the notifications according to physical state (packaging)

The analysis of notifications allowed us to verify that, in the great majority of cases, the physical state (packaging) of products was identified. Through analysis of Figure A2.11 it can be seen that there is a great preponderance (98%) of six

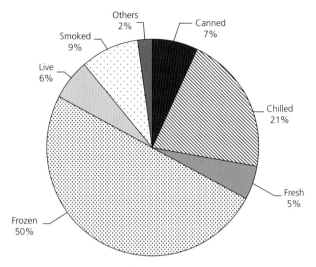

**Figure A2.11** Notifications for the main physical state (packaging).

physical states (packaging). The remaining 2% correspond to other physical states (packaging) and to notifications that did not possess identification. Figures A2.12–A2.17 allow detailed analysis of the notifications for each of the main six physical states (packaging).

### Frozen

Analysis of Figure A2.12 reveals that:

- mercury is the hazard with the highest number of reports in frozen fish (35.7%), followed by nitrofuran (23.2%) and cadmium (17.5%);
- since 2011 mercury notifications have been increasing, reaching its maximum at 66 in 2014;
- histamine reports have been relatively low and stable over the last 10 years; and
- as for the *Anisakis* spp. notifications, with exception of 2011 when there were 38 reports, the number of notifications has been relatively low and always less than 10 notifications per year.

### Chilled

From all notifications identified as the chilled physical state, Figure A2.13 shows the analysis of the five main hazards (76% of all occurrences).

- Mercury is the hazard with the highest number of reports with a total of 196 (36.4% of all notifications for the period considered), presenting an increasing tendency of reports over the last 10 years.
- The second most-reported hazard is *Anisakis* spp. (31.2%), although the number of notifications has lowered since its maximum value in 2011.

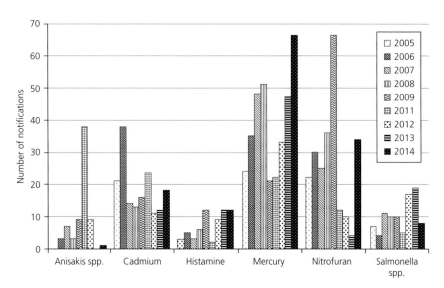

**Figure A2.12** Hazards notified in *frozen* physical state (packaging).

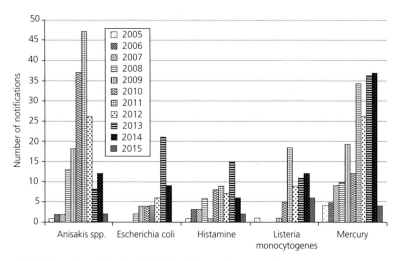

**Figure A2.13** Hazards notified in *chilled* physical state (packaging).

## Canned

Of all notifications registered to this physical state (packaging), 90% corresponded to four hazards: benzo(a)pyrene, cadmium, dioxins, and histamine (Fig. A2.14).

- Histamine had the biggest number of notifications during the time period studied (55.8%).
- Cadmium notifications have been steadily decreasing since 2006, having only one report registered in the last 3 years.

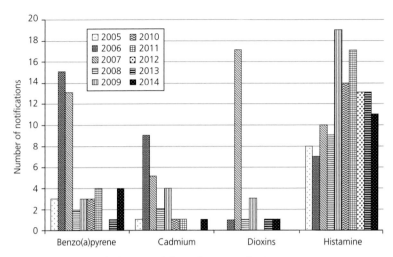

**Figure A2.14** Hazards notified in *canned* physical state (packaging).

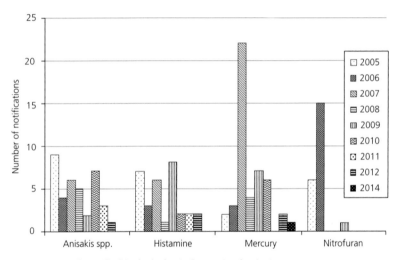

**Figure A2.15** Hazards notified in *fresh* physical state (packaging).

- Dioxins present a low number of notifications, except in 2007. In this year the notifications analysis highlighted that this abnormal value resulted from the contamination of a single product – cod – which represented 16 of the 17 registered reports.

## Fresh

Analysis of Figure A2.15 reveals the following.
- The number of notifications is globally low. The four identified hazards represent 2.4% of the total reports.

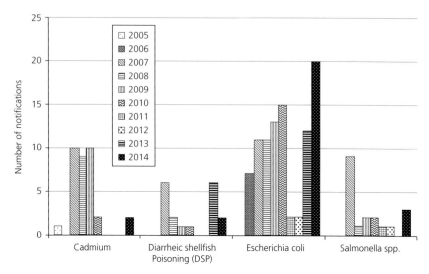

**Figure A2.16** Hazards notified in *live* physical state (packaging).

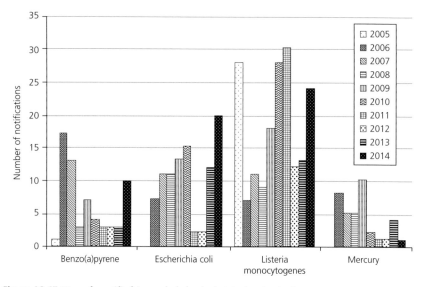

**Figure A2.17** Hazards notified in *smoked* physical state (packaging).

- Of the analysis of the notifications for mercury and for nitrofuran, 22 and 15 reports were registered in 2007 and 2006 and listed in Tables A2.4 and A2.5, respectively.

  According to Table A2.4, it is possible to conclude that the peak of notifications for mercury in 2007 was due to 20 reports in fresh swordfish. From Table A2.5 it

**Table A2.4** Mercury notifications in 2007 *fresh* physical state (packaging)

| Product | No. notifications |
|---|---|
| Grouper | 1 |
| Snapper | 1 |
| Swordfish | 20 |
| TOTAL | 22 |

**Table A2.5** Nitrofuran notifications in 2006 *fresh* physical state (packaging)

| Row labels | 2006 |
|---|---|
| Prawn | 4 |
| Sea bass | 1 |
| Shrimps | 10 |
| TOTAL | 15 |

is possible to conclude that shrimps were the primary source of the peak of nitro-furan notifications in 2006. Since these occurrences, the presence of nitrofurans was registered only once in the period up to the end of 2014.

## Live

Analysis of Figure 2.16 reveals the following.

- *E. coli* represents about 57% of all notifications.
- Cadmium remained stable for 3 years (9–10 notifications) between 2007 and 2009. In the subsequent years the number of reports was relatively low, even absent at some point.
- *Salmonella* spp. notifications have been low and stable since 2008 (ranging from 0 to 3 reports).

## Smoked

The three main hazards identified in smoked physical state (benzo(a)pyrene, *Listeria monocytogenes*, and mercury) represent about 92% of all notifications. Analysis of Figure A2.17 reveals the following.

- *Listeria monocytogenes* is the hazard with the largest number of reports, repre-senting 59% of its total.
- Although there was an increase in 2014, in the previous 6 years benzo(a)pyrene had a low number of notifications, ranging from 3 to 7 occurrences.
- The number of notifications for mercury has been consistently low since 2010.

Due to the fact that *Listeria monocytogenes* had a high percentage of notifications for this physical state (packaging), an evaluation of the most-reported products was performed. Smoked salmon is clearly the main product reported, with almost 70% of the total notifications (Fig. A2.18).

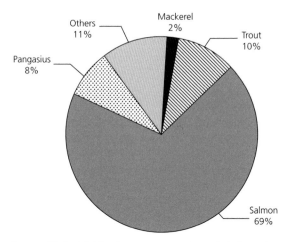

**Figure A2.18** Product notified with *Listeria monocytogenes* in *smoked* physical state (packaging).

## Analysis of the notifications according to product category

Analysis of Figure A2.19 reveals the following points.

- Until 2002, there only existed reports for fish and fish products, and crustaceans and products thereof.
- It was only in 2005 that an occurrence in cephalopods and products thereof was reported for the first time. After that the number of notifications remained relatively low and stable, representing 3.2% of the total reports.
- The number of occurrences in fish and fish products represents 59% of the total reports. There was a steady increase until 2004, but from that year on the number of notifications ranged from 215 to 330.
- Regarding bivalves and products thereof, the number of notifications increased significantly in the last 2 years, having almost tripled relative to the average of the previous 5 years (2008–2012).

    Analysis of Figure A2.20 reveals the following points.

- For the fish and fish products family, there was a significant increase in the number of notifications between 2000 and 2004 (625 reports) and between 2005 and 2014 (about 1338 reports).
- Notifications for the bivalves family doubled from the second to the third period of the study.
- Regarding crustaceans and products thereof, there was a decrease in the number of reports over the last 5 years of the analysis. The average of notifications from 2000 to 2009 is 628, while in the last period it dropped to 236 notifications.

    After a global analysis of the notifications for product category, results were observed in more detail. Figure A2.21 depicts the number of notifications of each hazard classification for each product category.

- Chemical hazards are the most reported in all product categories, with the exception of bivalve mollusks and products thereof, for which biological hazards reports are higher.

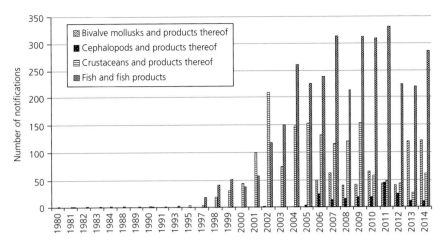

**Figure A2.19** Notifications by product category since 1980.

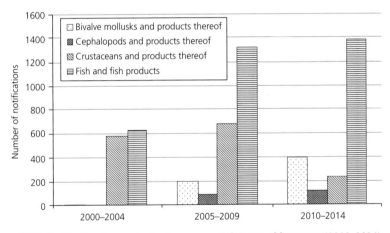

**Figure A2.20** Notifications by product category grouped in sets of five years (2000–2014).

- Reports regarding physical hazards represent only 0.8% of the total, the majority of which (>75%) occur in fish and fish products.

  Notifications were subsequently grouped by physical state (packaging) for each product category in the period 2005–2014. Again, only the six main physical states (packaging) were considered.

## Bivalve mollusk notifications

In this case study, there were no notifications in 2005. The smoked state did not have any notifications during the analyzed period either. The main conclusions from analysis of Figure A2.22 are as follows.

- There are reports for three main groups: chilled, frozen, and live.
- Canned and fresh states present an insignificant number of notifications during the period studied (they represent a combined 1.79% of all notifications).

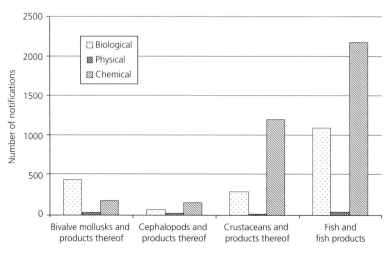

**Figure A2.21** Notifications grouped by hazard classification for each product category.

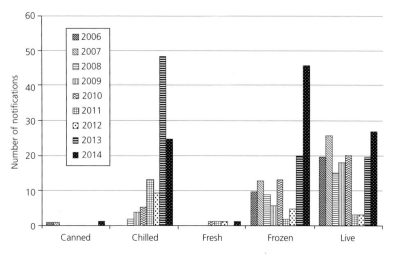

**Figure A2.22** Bivalve mollusks notifications by physical state (packaging).

• The number of notifications for the frozen state has doubled from 2013 to 2014, in contrast to the chilled state which in 2014 had about half of the notifications registered in 2013.

According to Figure A2.22, the number of notifications for the frozen state more than doubled from 2013 to 2014. In order to determine which hazard contributed to this increase, a more detailed analysis was performed by considering the food products in which the notifications occurred. According to Table A2.6, it is possible to verify that in 2014 the largest number of reports involved clams and the main hazard detected was norovirus (29 notifications).

**Table A2.6** Notifications for frozen bivalve mollusks

| Product/hazard | 2013 | 2014 | Total |
|---|---|---|---|
| Clams | | | |
| *Escherichia coli* | 2 | 2 | 4 |
| Hepatitis A virus | 0 | 1 | 1 |
| Norovirus | 0 | 24 | 24 |
| *Salmonella* spp. | 17 | 5 | 22 |
| Clams total | 19 | 32 | 51 |
| Mussels | | | |
| Cadmium | 0 | 5 | 5 |
| Diarrheic shellfish poisoning (DSP) | 1 | 1 | 2 |
| Norovirus | 0 | 5 | 5 |
| Mussels total | 1 | 11 | 12 |
| Scallops | | | |
| *Clostridium* spp. | 0 | 1 | 1 |
| Scallops total | 0 | 1 | 1 |
| Shrimp | | | |
| Oxytetracycline | 0 | 1 | 1 |
| Shrimps total | 0 | 1 | 1 |
| Squids | | | |
| Cadmium | 0 | 1 | 1 |
| Squids total | 0 | 1 | 1 |
| TOTAL | 20 | 46 | 66 |

In the case of chilled state, from 2013 to 2014 the notifications almost reduced by half. The last two years were analyzed in order to understand which food product demonstrated a decrease in the number of notifications and which hazards were involved. According to Table A2.7 clams are the food product with the highest number of notifications in both 2013 and in 2014. As for the hazards, those whose occurrence decreased were *E. coli* (12 less reports in 2014) and norovirus (9 less reports in 2014).

## Cephalopod notifications

The main conclusions from analysis of Figure A2.23 can be found as:
- it is clear that the frozen state dominates the number of notifications (84%);
- in this case, the fresh and live states have no notifications; and
- smoked and chilled states represent a single notification each.

## Crustacean notifications

The main conclusions from analysis of Figure A2.24 are as follows.
- The state with the largest number of notifications is the frozen state, representing 81.9% of all notifications. In this state the average number of notifications in the first 5 years is 77, having decreased significantly in the last 5 years to 31.

**Table A2.7** Notifications for chilled bivalve mollusks

| Product/hazard | 2013 | 2014 | Total |
|---|---|---|---|
| Clams | | | |
| Cadmium | 1 | 0 | 1 |
| *Escherichia coli* | 17 | 5 | 22 |
| Norovirus | 10 | 7 | 17 |
| *Vibrio* spp. | 2 | 0 | 2 |
| Clams total | 30 | 12 | 42 |
| Mussels | | | |
| Azaspiracid shellfish poisoning (AZP) | 3 | 0 | 3 |
| Diarrheic shellfish poisoning (DSP) | 0 | 2 | 2 |
| *Escherichia coli* | 4 | 4 | 8 |
| Norovirus | 0 | 1 | 1 |
| *Salmonella spp.* | 1 | 0 | 1 |
| Mussels total | 8 | 7 | 15 |
| Oysters | | | |
| Norovirus | 10 | 3 | 13 |
| Oysters total | 10 | 3 | 13 |
| Scallops | | | |
| Amnesic shellfish poisoning | 1 | 1 | 2 |
| E451 - triphosphate | 0 | 1 | 1 |
| Paralytic shellfish poisoning (PSP) | 0 | 1 | 1 |
| Scallops total | 1 | 3 | 4 |
| TOTAL | 49 | 25 | 74 |

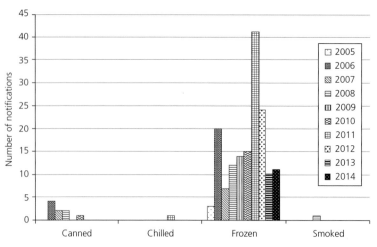

**Figure A2.23** Cephalopods notifications by physical state (packaging).

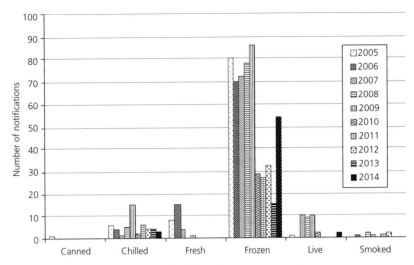

**Figure A2.24** Crustaceans notifications by physical state (packaging).

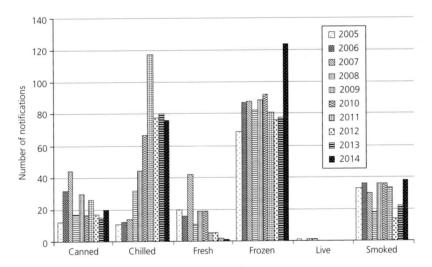

**Figure A2.25** Fish notifications by physical state (packaging).

**Table A2.8** Notifications for frozen fish (2013–2014)

| Row Labels | 2013 | 2014 |
|---|---|---|
| *Listeria monocytogenes* | 2 | 3 |
| Mercury | 47 | 66 |
| Nitrofuran | 0 | 20 |
| Egg (Allergen) | 1 | 7 |

- All other states have a much lower number of notifications, especially the canned state which was notified only once and the fresh state which, since 2007, only registered one notification.

## Fish and fish products notifications

The main conclusions from analysis of Figure A2.25 include the following.

- The frozen state is the one with the highest number of notifications (866). This number was relatively stable during the whole period studied except in 2014, when there was an increase of about 30% compared to 2013.
- In the fresh state there has been a steady decrease in the number of notifications.
- The live state has only three notifications in total and none since 2008.
- The number of reports for smoked and canned states has been relatively constant.

Figure A2.25 demonstrates a significant increase in the frozen state during 2013–2014. In order to understand which hazard caused this variation, a more detailed analysis was conducted. Due to the fact that the list of hazards is very extensive, they are not be all mentioned; instead, Table A2.8 presents the main hazards responsible for the 2014 outbreak.

After analyzing Table A2.8 it is possible to conclude that in 2013 there were no notifications for nitrofurans, but they increased to 20 in 2014. Mercury and *Listeria monocytogenes* are also responsible for the significant increase of the number of reports (16 and 18 respectively).

# References

Aberoumand, A. (2010) The effect of water activity on preservation quality of fish, a review article. *World Journal of Fish and Marine Sciences*, **2**(3), 221–225.

Adachi, K. & Hirata, T. (2011) Blackening of crustaceans during storage: mechanism and prevention. In *Handbook of Seafood Quality, Safety and Health Applications* (eds C. Alasalvar, K. Mitashita, F. Shahidi & U. Wunasundara). Oxford: John Wiley & Sons, Ltd, pp. 109–118.

Ahmed, F.E. (1991) *Seafood Safety*. Virginia: National Academy Press.

Aitken, A., Lees, A. & Smith, J.G.M. (2001) *Measuring Fish Composition*. Torry Research Station: Food and Agriculture Organization of the United Nations.

Anon (2014) PAS 96:2014 Guide to protecting and defending food and drink from deliberate attack.

Anon (2015) Risk Ranger: Food Safety Centre. Available at http://www.foodsafetycentre.com.au/riskranger.php (accessed 10 September 2015).

APHA & FDA (1972) *Proceedings of the 1971 National Conference on Food Protection*. Washington, DC: US Government Printing Office.

Arfini, F. & Mancini, M.C. (2014) British Retail Consortium (BRC) Standard: a New Challenge for Firms Involved in the Food Chain. Analysis of Economic and Managerial Aspects, p. 9. Institute of Agrifood Economics, Department of Economic and Quantitative Studies, Parma University.

Auerbach, P.S. (2011) *Wilderness Medicine: Expert Consult Premium Edition – Enhanced Online Features*. Sixth edition. Missouri: Elsevier Health Sciences.

Bae, J.H., Hwang, S.Y., Yoon, S.H., Noh, I. & Lium, S.Y. (2011) Comparison between ordinary and dark muscle extracts of yellowtail (*Serila quinqueradiata*) on chemical characteristics, antiproliferative and antioxidant properties. *Journal of Food Technology*, **9**(3), 99–105.

Bohl, M., Bach, P. & Bartmann, K. (1999) *Zucht und Produktion von Süsswasserfischen* Second edition. Frankfurt: DLG-Verlag.

Boyle, P. & Rodhouse, P. (2005) *Cephalopods: Ecology and Fisheries*. Oxford: Wiley-Blackwell.

BRC (2015) *Global Standards for Food Safety, Issue 7*. London: BRC, 118 pp.

BSI (2008a) BSI British Standards publishes food safety specification for the food manufacturing industry. Available at: http://www.bsigroup.com/en-GB/about-bsi/media-centre/press-releases/2008/11/BSI-British-Standards-publishes-food-safety-specification-for-the-food-manufacturing-industry/#.VTYreSFViko (accessed 3 September 2015).

BSI (2008b) PAS 220:2008 Prerequisite programmes on food safety for food manufacturing. 17 pp.

Buchanan, R.L. & Whiting, R.C. (1998) Risk assessment: A means for linking HACCP plans and public health. *Journal of Food Protection*, **61**, 531–1534.

CAC (1969) CAC/RCP 1-1969 v.4 – 2003 General Principles of Food Hygiene. Codex Alimentarius, p. 31.

CAC (1979) CAC/RPC 23-1979 v.2 – 1993 Code of Hygienic Practice for Low-Acid and Acidified Low-Acid Canned Foods, 85 pp.

*Food Safety in the Seafood Industry: A Practical Guide for ISO 22000 and FSSC 22000 Implementation,*
First Edition. Nuno F. Soares, Cristina M. A. Martins and António A. Vicente.
© 2016 John Wiley & Sons, Ltd. Published 2016 by John Wiley & Sons, Ltd.

CAC (1999) CAC/GL 30 Principles and guidelines for the application of microbiological risk assessment. CX/FH 96/10. August, 1996. www.codexalimentarius.net/download/standards/357/CXG_030e.pdf.

CAC (2003) CAC/RCP 52-2003 rev. 2013 Code of Practice for Fish and Fishery Products, 238 pp.

CAC (2010) General Standard for the Labelling of Prepackaged Foods, p.7. Available at: http://www.codexalimentarius.org/download/standards/32/CXS_001e.pdf (accessed 3 September 2015).

CAC (2011) FFP/31 Thailand information paper on estimating the risk of developing histamine poisoning from the consumption Thai fish sauces. FFP/31 CRD 18. Available at: ftp://ftp.fao.org/codex/meetings/CCFFP/ccffp31/CRD/CRD_18_Thailand.pdf (accessed 10 September 2015).

Chaijan, M., Benjakul, S., Visessanguan, W. & Fuastman, C. (2004) Characteristics and gel properties of muscles from sardine (Sardinella gibbosa) and mackerel (Rastrelliger kanagurta) caught in Thailand. *Food Research International*, **37**, 1021–1030.

Clever, J., Rasdall, B. & Jie, M. (2015) GB14881 China - Peoples Republic of China's General Hygiene Regulation for Food Production.

Coan, E.V. & Valentich-Scott, P. (2006) Marine Bivalvia. In *The Mollusks: A Guide to Their Study, Collection, and Preservation* (eds C.F. Sturm, T.A. Pearce & Á. Valdés). Pittsburgh, PA: American Malacological Society, 339–347.

Cohen, J.T., Bellinger, D.C. *et al.* (2005) A quantitative risk-benefit analysis of changes in population fish consumption. *American Journal of Preventive Medicine*, **29**(4), 325–334.

Crawford, M.A., Bloom, M. *et al.* (1999) Evidence for the unique function of docosahexaenoic acid during the evolution of the modern hominid brain. *Lipids*, **34**(Suppl), S39–S47.

Cunnane, S.C. & Crawford, M.A. (2003) Survival of the fattest: Fat babies were the key to evolution of the large human brain. *Comparative Biochemistry and Physiology - A Molecular and Integrative Physiology*, **136**(03), 17–26.

Daul, C.B., Slattery, M., Reese, G. & Lehrer, S.B. (1994) Identification of the major brown shrimp (*Penaeus aztecus*) allergen as the muscle protein tropomyosin. *International Archives of Allergy and Immunology*, **105**(1), 49–55.

Dauthy, M.E. (1995) *Fruit and Vegetable Processing*. 119th edition. Rome: Food and Agriculture Organization of the United Nations.

Davies, J.H.V. (1970) The Codex Alimentarius. *Journal of the Association of Public Analysts*, **8**, 53–67.

de Benoist, B., McLean, E., Andersson, M. & Rogers, L. (2008) Iodine deficiency in 2007: Global progress since 2003. *Food and Nutrition Bulletin*, **29**(3), 195–202.

Delgado, C. (2006) NP EN ISO 22000 Implementação de Sistemas de Gestão e Segurança Alimentar, 189 pp.

Dovydaitis, T. (2008) Fish consumption during pregnancy: an overview of the risks and benefits. *Journal of Midwifery & Women's Health*, **53**(4), 325–330.

Dunning, M. & Wadley, V.A. (1998) *Cephalopods of Commercial Importance in Australian Fisheries*. Melbourne: CSIRO Publishing.

EC (2005) Implementation of procedures based on the HACCP principles , and facilitation of the implementation of the HACCP principles in certain food businesses, 26 pp. Available at: http://ec.europa.eu/food/food/biosafety/hygienelegislation/guidance_doc_haccp_en.pdf (accessed 20 September 2015).

EC (2007) Commission Regulation (EC) No 1441/2007 of 5 December 2007 amending Regulation (EC) No 2073/2005 on microbiological criteria for foodstuffs. *Official Journal of the European Union*, **50**, 12–29.

EC (2010) Commission Regulation (EU) No 37/2010 of 22 December 2009 on pharmacologically active substances and their classification regarding maximum residue limits in foodstuffs of animal origin. *Official Journal of the European Union*, **53**, 1–72.

EC (2011) MEMO/11/783 Questions and answers on food additives. Available at: http://europa.eu/rapid/press-release_MEMO-11-783_en.htm (accessed 20 September 2015).

EC & EP (2002) Directive 2001/95/EC of the European Parliament and of the Council of 3 December 2001 on general product safety. *Official Journal of the European Communities*, **45**, 4–17.

EC & EP (2004) Regulation (EC) No. 853/2004 of the European Parliament and of the Council of 29 April 2004 laying down specific hygiene rules for on the hygiene of foodstuffs. *Official Journal of the European Union*, **47**, 55–205.

EC & EP (2011) Regulation (EU) No. 1169/2011 of the European Parliament and of the Council of 25 October 2011 on the provision of food information to consumers, amending Regulations (EC) No 1924/2006 and (EC) No 1925/2006 of the European Parliament and of the Council. *Official Journal of the European Union*, **54**, 18–63.

EFSA (2005) Opinion of the scientific panel on contaminants in the food chain on a request from the european Parliament related to the safety assessment of wild and farmed fish. *The EFSA Journal*, **236**, 1–118.

EFSA (2008) Findings of the EFSA Data Collection on Polycyclic Aromatic Hydrocarbons in Food. Available at: http://www.efsa.europa.eu/sites/default/files/scientific_output/files/main_documents/datex_report_update_en%2C0.pdf (accessed 20 September 2015).

EFSA (2010) Results of the monitoring of dioxin levels in food and feed. *EFSA Journal*, **8**(3), 1385.

FAO (1960) *Report of the Conference for Europe, 10–15 October 1960.* Rome: Food and Agriculture Organization of the United Nations.

FAO (1962) *Resolution No. 12/61. Report of the Eleventh Session of the Conference, 4–24 November 1961.* Rome: Food and Agriculture Organization of the United Nations.

FAO (1998) *Food Quality and Safety Systems - A Training Manual on Food Hygiene and the Hazard Analysis and Critical Control Point (HACCP) System.* Rome: Food and Agriculture Organization of the United Nations.

FAO (1999) Expert consultation on the trade impact of *Listeria* in fish products. Amherst, 17–20 May 1999. *FAO Fisheries Report* No. 604. Food and Agriculture Organization of the United Nations, Viale delle Terme di Caracalla, 00100 Rome, Italy.

FAO (2004) Application of risk assessment in the fish industry. FAO Fisheries Technical Paper 442. Food and Agriculture Organization of the United Nations, Viale delle Terme di Caracalla, 00100 Rome, Italy. Available at ftp://ftp.fao.org/docrep/fao/007/y4722e/y4722e00.pdf (accessed 10 September 2015).

FAO (2012) *The State of World Fisheries and Aquaculture 2012.* Rome: Food and Agriculture Organization of the United Nations.

FAO (2013) *Fish Identification Tools for Biodiversity and Fishereis Assessments* (ed. J. Fischer). Rome: Food and Agriculture Organization of the United Nations.

FAO (2014) *The State of World Fisheries and Aquaculture 2014.* Rome: Food and Agriculture Organization of the United Nations.

FAO & WHO (2006a) Food safety risk analysis: A guide for national food safety authorities. FAO and Food Nutrition Paper no. 87.

FAO & WHO (2006b) *Understanding the Codex Alimentarius Third.* Rome: Food and Agriculture Organization of the United Nations.

FAO & WHO (2009) *Food Hygiene: Basic Texts*, 4th edition. Rome: Food and Agriculture Organization of the United Nations.

FAO & WHO (2010) CAC/RCP 53-2003 Code of Hygienic Practice for Fresh Fruits and Vegetables, 28 pp.

FAO & WHO (2011) *Report of the Joint FAO/WHO Expert Consultation on the Risks and Benefits of Fish Consumption.* Geneva: World Health Organisation.

FAO & WHO (2014) CODEX STAN 192-1995 General standard for food additives.

FAO & WHO (2015) *Codex Alimentarius Commission Procedural Manual.* Rome: Food and Agriculture Organization of the United Nations.

FDA (2011) *Fish and Fishery Products: Hazards and Controls Guidance*. Fourth edition (ed. B. Leonard). Darby, PA: Diane Publishing Company.

FDA (2012a) *Bad Bug Book: Handbook of Foodborne Pathogenic Microorganisms and Natural Toxins*. Available at: http://www.fda.gov/Food/FoodborneIllnessContaminants/CausesOfIllnessBadBugBook/ (accessed 20 September 2015).

FDA (2012b) Code of Federal Regulations. Title 21: Food and Drugs. Available at: http://www.gpo.gov/fdsys/pkg/CFR-2012-title21-vol1/pdf/CFR-2012-title21-vol1.pdf (accessed 20 September 2015).

FDA (2013) Food Code - 2013 Recommendations of the United States Public Health Service Food and Drug Administration. Available at: http://www.fda.gov/downloads/Food/GuidanceRegulation/RetailFoodProtection/FoodCode/UCM374510.pdf (accessed 20 September 2015).

FDA (2014) Safe Food Handling: What You Need to Know. US Food and Drug Administration: *Protecting and Promoting Your Health*. Available at: http://www.fda.gov/food/resourcesforyou/consumers/ucm255180.htm#chart (accessed 3 September 2015).

Feeley, R.M., Criner, P.E. & Watt, B.K. (1972) Cholesterol content of foods. *Journal of American Dietetic Association*, **61**(2), 134–149.

FFSC (2013) FSSC 22000 Guidance on Appendix IA. Available at: http://www.fssc22000.com/documents/pdf/guidances/guidance-on-appendix-ia.pdf (accessed 20 September 2015).

FFSC (2014) FSSC 22000 Guidance document on certification scopes. Available at: http://fssc22000.com/downloads/FSSCscopeguidancevs5.pdf (accessed 20 September 2015).

FFSC (2015a) FSSC 22000 Certificates Directory. Available at: http://www.fssc22000.com/documents/certified-organizations.xml?lang=en (accessed 20 September 2015).

FFSC (2015b) FSSC 22000 Certification scheme for food safety systems in compliance with ISO 22000:2005 and technical specifications for sector PRPs: Part I: Requirements for organizations that require certification. Available at: http://www.fssc22000.com/documents/pdf/certification-scheme/fssc22000_part1-v3.1_2014.pdf (accessed 20 September 2015).

FFSC (2015c) Legal Status. Food Safety System Certification 22000, 2 pp. Available at: http://www.fssc22000.com/documents/about-us/legal-status.xml?lang=en (accessed 3 September 2015).

Garcia, A.M., Vieira, J.P., Winemiller, O. & Grimm, A.M. (2004) Comparison of 1982–1983 and 1997–1998 El Niño effects on the shallow-water fish assemblage of the Patos Lagoon estuary (Brazil). *Estuaries*, **27**(6), 905–914.

Garcia, S.M., Zerbi, A. *et al.* (2003) The ecosystem approach to fisheries. *FAO Fisheries Technical Paper*, 443,71 pp. Available at: ftp://ftp.fao.org/docrep/fao/006/y4773e/y4773e00.pdf (accessed 3 September 2015).

GFSI (2013) GFSI Guidance Document, sixth edition, issue 3, version 6.3. Available at http://www.mygfsi.com/schemes-certification/benchmarking/gfsi-guidance-document.html (accessed 18 September 2015).

GFSI (2015) Recognised Schemes. *MyGFSI*. Available at: http://www.mygfsi.com/schemes-certification /recognised-schemes.html (accessed 3 September 2015).

GFSI & Sealed Air (2014) *GFSI Efficacy Research Project*. Leeds: McCallum Layton.

Gosling, E. (2003) *Bivalves Molluscs: Biology, Ecology and Culture*. Oxford: John Wiley & Sons, Ltd.

Granata, L.A., Flick, G.J. & Martin, R.E. (eds) (2012) Shellfish: mollusks. In *The Seafood Industry: Species, Products, Processing, and Safety*, second edition. Wiley-Blackwell.

Haszprunar, G. (2001) Mollusca (Molluscs). In *Encyclopedia of Life Sciences*. Chichester: John Wiley & Sons, Ltd.

Helm, M.M., Bourne, N. & Lovatelli, A. (2004) *Hatchery Culture of Bivalves: A Practical Manual*. Rome: Food and Agriculture Organization of the United Nations.

Helm, R.M. & Burks, A.W. (2009) Food Allergens. In *Allergy and Allergic Diseases*, Volume 1 (ed. A. B. Kay). Wiley-Blackwell, 1146–1163.

Huss, H.H. (1988) *Fresh Fish - Quality and Quality Changes: A Training Manual Prepared for the FAO/DANIDA Training Programme on Fish Technology and Quality Control*. Rome: Food and Agriculture Organization of the United Nations.

Huss, H.H. (1995) *Quality and Quality Changes in Fresh Fish*. Rome: Food and Agriculture Organization of the United Nations.

Huss, H.H., Ababouch, L. & Gram, L. (2004) Assessment and management of seafood safety and quality. FAO Fisheries Technical Paper no. 444, 230 pp.

IFS (2014) IFS Food (Version 6). Standard for auditing quality and food safety of food products. IFS International Featured Standards, April, p. 149. Available at: http://www.ifs-certification.com/index.php/en/retailers-en/ifs-standards/ifs-food (accessed 3 September 2015).

IFS (2015) History of IFS. *International Featured Standards*. Available at: http://www.ifs-certification.com/index.php/en/ifs-certified-companies-en/introduction-to-ifs/ifs-history (accessed 5 September 2015).

Ishikawa, K. (1986) *Guide to Quality Control*. Tokyo: Asian Productivity Organization.

ISO (2005a) ISO 22000:2005 Food safety management systems: Requirements for any organization in the food chain, 32 pp. Available at: http://www.iso.org/iso/catalogue_detail?csnumber=35466 (accessed 20 September 2015).

ISO (2005b) ISO 9000:2005 Quality management systems: Fundamentals and vocabulary, 30 pp. Available at: http://www.iso.org/iso/catalogue_detail?csnumber=42180 (accessed 20 September 2015).

ISO (2005c) ISO/TS 22004:2005 Food safety management systems: Guidance on the application of ISO 22000:2005. Available at: http://www.iso.org/iso/home/store/catalogue_tc/catalogue_detail.htm?csnumber=39835 (accessed 20 September 2015).

ISO (2007) ISO 22005:2007 Traceability in the feed and food chain: General principles and basic requirements for system design and implementation. Available at: http://www.iso.org/iso/catalogue_detail?csnumber=36297 (accessed 20 September 2015).

ISO (2008a) ISO 22000 Food Safety. *ISO Management Systems*, **8**(3), 53.

ISO (2008b) ISO 9001:2008 Quality management systems: Requirements. Available at: http://www.iso.org/iso/home/store/catalogue_tc/catalogue_detail.htm?csnumber=46486 (accessed 20 September 2015).

ISO (2011) ISO 19011:2011 Guidelines for auditing managements systems. Available at: http://www.iso.org/iso/home/store/catalogue_tc/catalogue_detail.htm?csnumber=50675 (accessed 20 September 2015).

ISO (2014) ISO 22004:2014 Food safety management systems: Guidance on the application of ISO 22000:2005. Available at http://www.iso.org/iso/catalogue_detail?csnumber=60992 (accessed 20 September 2015).

James, D. (2013) Risks and Benefits of Seafood Consumption. Globefish Research Programme, Report no. **108**, 28 pp.

Jay, J.M., Loessner, M.J. & Golden, D.A. (2005) *Modern Food Microbiology*, seventh edition. New York: Springer Science + Business Media.

Jereb, P. & Roper, C.F.E. (2005) *Cephalopods of the World, Volume 1*. Rome: Food and Agriculture Organization of the United Nations.

Johnston, W.A., Nicholson, F.J., Roger, A. & Stroud, G.D. (1994) Freezing and Refrigerated Storage in Fisheries. FAO Fisheries Technical Paper, 340.

Kobayashi, A., Tanaka, H. *et al.* (2006) Comparison of allergenicity and allergens between fish white and dark muscles. *Allergy: European Journal of Allergy and Clinical Immunology*, **61**(3), 357–363.

Lewbart, G.A. (ed.) (2011). Crustaceans. In *Invertebrate Medicine*, second edition. Wiley-Blackwell.

Lopata, A.L., O'Hehir, R.E. & Lehrer, S.B. (2010) Shellfish allergy. *Clinical and Experimental Allergy*, **40**, 850–858.

Love, R.M. (1970) *The Chemical Biology of Fishes*. London: Academic Press.

Love, R.M. (1980) *The Chemical Biology of Fishes*. Volume 2, Advances 1968–1977 with a supplementary key to the chemical literature. London: Academic Press.

Love, R.M. (1988) *The Food Fishes: Their Intrinsic Variation and Practical Implications*. London: Farrand Press.

Martin, R.E., Carter, E.P., Flick, G.J. & Davis, L.M. (eds) (2000) *Marine & Freshwater Products Handbook*. Lancaster, Pennsylvania: Technomic Publishing Company, Inc.

Mensah, L.D. & Julien, D. (2011) Implementation of food safety management systems in the UK. *Food Control*, **22**(8), 1216–1225.

Mozaffarian, D. & Rimm, E.B. (2006) Fish intake, contaminants, and human health. *JAMA: Journal of the American Medical Association*, **296**(15), 1885–1899.

Murray, J. & Burt, J.R. (2001) The Composition of Fish. Torry Advisory Note No. 38, Torry Research Station.

Nelson, J.S. (2006) *Fishes of the World*. Fourth edition. Hoboken, New Jersey: John Wiley & Sons, Inc.

Neumeyer, K., Ross, T., Thomson, G. & McMeekin, T.A. (1997) Validation of a model describing the effects of temperature and water activity on the growth of psychrotrophic pseudomonads. *International Journal of Food Microbiology*, **38**, 55–63.

Ng, P.K.L. (1998) Crabs. In *The Living Marine Resources of the Western Central Pacific* (eds K.E. Carpenter & V.H. Niem). Rome: Food and Agriculture Organization of the United Nations, 1045–1055.

Notermans, S. & Mead, G.C. (1996) Incorporation of elements of quantitative risk analysis in the HACCP system. *International Journal of Food Microbiology*, **30**, 157–173.

OECD & FAO (2013) *OECD-FAO Agricultural Outlook 2013–2022*. Available at: http://www.oecd-ilibrary.org/agriculture-and-food/oecd-fao-agricultural-outlook-2013_agr_outlook-2013-en (Accessed 20 September 2015).

OECD & FAO (2014) *OECD-FAO Agricultural Outlook 2014–2023*. Available at: http://www.oecd-ilibrary.org/agriculture-and-food/oecd-fao-agricultural-outlook-2014_agr_outlook-2014-en (accessed 20 September 2015).

Osman, H., Suriah, A.R. & Law, E.C. (2001) Fatty acid composition and cholesterol content of selected marine fish in Malaysian waters. *Food Chemistry*, **73**(1), 55–60.

Pirestani, S., Ali Sahari, M., Barzegar, M. & Seyfabai, S.J. (2009) Chemical compositions and minerals of some commercially important fish species from the South Caspian Sea. *International Food Research Journal*, **16**(1), 39–44.

Poutiers, J.M. (1998) Bivalves (Acephala, Lamellibranchia, Pelecypoda). In *The Living Marine Resources of the Western Central Pacific* (eds K.E. Carpenter & V.H. Niem). Rome: Food and Agriculture Organization of the United Nations, 123–141.

Ralston, N.V., Ralston, C.R., Blackwell, J.L. & Raymond, L.J. (2008) Dietary and tissue selenium in relation to methylmercury toxicity. *Neurotoxicology*, **29**(5), 802–811.

Raven, P.H. & Johnson, G.B. (2002) *Biology*, sixth edition. Boston: McGraw-Hill, 872 pp.

Ray, A.K. (2008) *Fossils in Earth Sciences*. New Delhi: PHI Learning Pvt. Ltd.

Roberts, R.J. (ed.) (2012) The anatomy and physiology of teleosts. In *Fish Pathology*, Fourth edition. Wiley-Blackwell.

Ross, T. & Sumner, J.L. (2002) A simple, spreadsheet-based, food safety risk assessment tool. *International Journal of Food Microbiology*, **77**, 39–53.

Ross-Nazzal, J. (2007) 'From Farm to Fork': How Space Food Standards Impacted the Food Industry and Changed Food Safety Standards. In *Societal Impact of Space Flight* (eds S.J. Dick & R.D. Launius). Washington, DC: NASA, 219–236.

Ryder, J., Iddya, K. & Ababouch, L. (2014) Assessment and management of seafood safety and quality: Current practices and emerging issues. FAO Fisheries and Aquaculture Technical Paper no. 574, 432 pp.

Saxena, A. (2005) *Text Book of Crustacea.* Darya Ganj, New Delhi: Discovery Publishing House.

Schultz, K. (2004) *Ken Schultz's Field Guide to Saltwater Fish.* Hoboken, New Jersey: John Wiley & Sons, Inc.

Seafish (2012) SR654: Review of polyphosphates as additives and testing methods for them in scallops and prawns. Campden BRI Report no. BC-REP-1258461-01.

Shanti, K.N., Martin, B.M. *et al.* (1993) Identification of tropomyosin as the major shrimp allergen and characterization of its IgE-binding epitopes. *International Archives of Allergy and Immunology,* **151**(10), 5354–5363.

Sidhu, K.S. (2003) Health benefits and potential risks related to consumption of fish or fish oil. *Regulatory Toxicology and Pharmacology,* **38**(3), 336–344.

Sperber, W.H. (2005) HACCP does not work from farm to table. *Food Control,* **16**(6), 511–514.

SQF (2008) SQF 2000 Code. A HACCP-based Supplier Assurance Code for the Food Manufacturing and Distributing Industries (6th Edition), 78 pp. Available at: https://www.sqfi.com/wp-content/uploads/SQF-2000-Code.pdf (accessed 20 September 2015).

SQF (2010) SQF 1000 Code. A HACCP-based Supplier Assurance Code for the Primary Producer (5th Edition), 60 pp. Available at: https://www.sqfi.com/wp-content/uploads/SQF-1000-Code.pdf (accessed 20 September 2015).

SQF (2014a) General Guidance for Developing, Documenting, Implementing, Maintaining, and Auditing an SQF System. Module 11: Good Manufacturing Practices for Processing of Food Products. Available at: http://www.sqfi.com/wp-content/uploads/SQF-Code-Ed.-7.1-Module-11-Guidance-Document.pdf (accessed 5 Sepetmber 2015).

SQF (2014b) SQF Code. A HACCP-based Supplier Assurance Code for the Food Industry (Edition 7.2). Available at: http://www.sqfi.com/wp-content/uploads/SQF-Code_Ed-7.2-July.pdf (accessed 20 September 2015).

Stansby, M.E. & Hall, A.S. (1967) Chemical composition of commercially important fish of the USA. *Fishery Industrial Research,* **3**(4), 29–34.

Toppe, J. (2012) Eat more fish - a healthy alternative. Farmed fish - a good choice. *FAN - FAO Aquaculture Newsletter,* **49**, 192.

Torpy, J.M. (2006) Eating fish: Health benefits and risks. *Journal of the American Medical Association,* **296**(15), doi:10.1001/jama.296.15.1926.

Tudhope, A.W., Chilcott, C.P. *et al.* (2001) Variability in the El Niño-Southern Oscillation through a glacial-interglacial cycle. *Science,* **291**(5508), 1511–1517.

UN (2015) *World Population Prospects: The 2012 Revision (DVD Edition).* Available at: http://esa.un.org/unpd/wpp/Publications/Files/Key_Findings_WPP_2015.pdf (accessed 20 September 2015).

USFDA (2015) FDA Food Safety Modernization Act (FSMA). http://www.fda.gov/Food/GuidanceRegulation/FSMA/default.htm.

USFDA/CFSAN (2015) FDA-iRISK version 2.0. FDA CFSAN. College Park, Maryland. Available at https://irisk.foodrisk.org/ (accessed 10 September 2015).

Wallace, C.A., Sperber, W.H. & Mortimore, S.E. (2010) *Food Safety for the 21st Century.* Oxford, UK: Wiley-Blackwell.

Wallace, C.A., Sperber, W.H. & Mortimore, S.E. (2011) *Food Safety for the 21st Century: Managing HACCP and Food Safety Throughout the Global Supply Chain.* Oxford, UK: Wiley-Blackwell.

Warner, K., Timme, W., Lowell, B. & Hirshfield, M. (2013) Ocean study reveals seafood fraud nationwide. Available at: http://oceana.org/sites/default/files/reports/National_Seafood_Fraud_Testing_Results_FINAL.pdf (accessed 20 September 2015).

Wheeler, A. & Jones, A.K.G. (1989) *Fishes Illustrated.* Cambridge: Cambridge University Press.

WHO (2006) Five Keys to Safer Food Manual. Available at: http://www.who.int/foodsafety/publications/consumer/manual_keys.pdf (accessed 5 September 2015).

WHO (2007) Exposure to Mercury: A Major Public Health Concern. Preventing Disease Through Healthy Environments. Available at: http://www.who.int/ipcs/features/mercury.pdf (accessed 5 September 2015).

WHO (2010a) Exposure to Arsenic: A Major Public Health Concern. Preventing Disease Through Healthy Environments. Available at: http://www.who.int/ipcs/features/arsenic.pdf (accessed 5 September 2015).

WHO (2010b) Exposure to Cadmium: a Major Public Health Concern. Preventing Disease Through Healthy Environments. Available at: http://www.who.int/ipcs/features/cadmium.pdf (accessed 5 September 2015).

WHO (2010c) Exposure to Dioxins and Dioxin-like Substances: a Major Public Health Concern. Preventing Disease Through Healthy Environments. Available at: http://www.who.int/ipcs/features/dioxins.pdf (accessed 5 September 2015).

WHO (2010d) Exposure To Highly Hazardous Pesticides: a Major Public Health Concern. Preventing Disease Through Healthy Environments. Available at: http://www.who.int/ipcs/features/hazardous_pesticides.pdf (accessed 5 September 2015).

WHO (2010e) Exposure to Lead: A Major Public Health Concern. Preventing Disease Through Healthy Environments. Available at: http://www.who.int/ipcs/features/lead.pdf (accessed 5 September 2015).

# Index

---

*Food Safety in the Seafood Industry: A Practical Guide for ISO 22000 and FSSC 22000 Implementation*,
First Edition. Nuno F. Soares, Cristina M. A. Martins and António A. Vicente.
© 2016 John Wiley & Sons, Ltd. Published 2016 by John Wiley & Sons, Ltd.